ACCIDENT AND DESIGN

ACCIDENT AND DESIGN

contemporary debates
in risk management

Edited by
Christopher Hood & David K. C. Jones
London School of Economics and Political Science

Routledge
Taylor & Francis Group

LONDON AND NEW YORK

First published in 1996 by UCL Press
Reprinted 2001 by
UCL Press Limited
2 Park Square, Milton Park,
Abingdon, Oxon, OX14 4RN

Transferred to Digital Printing 2005

UCL Press is an imprint of the Taylor & Francis Group

The name of University College London (UCL) is a registered
trade mark used by UCL Press with the consent of the owner.

British Library Cataloguing-in-Publication Data
A CIP catalogue record for this book is available from the British Library.

Library of Congress Cataloging-in-Publication Data are available

ISBNs: 1-85728-597-2 HB
       1-85728-598-0 PB

Typeset in Zapf Elliptical.

CONTENTS

CHAPTER FIVE
Designing institutions: a house of cards?   111

THE FEASIBILITY OF INSTITUTIONAL DESIGN IN RISK MANAGEMENT   111

PREFACE

"Four chapters good, two chapters bad" (with apologies to George Orwell) appears to have been orthodoxy's response to the 1992 [British] Royal Society publication, *Risk: analysis, perception and mana ement*. The "good" four chapters were those written by distinguished engineers, statisticians and natural scientists, which reflected a view of public risk management as properly the domain of science and engineering rather than of politics and economics. That is, both risk, and human behaviour in relation to risks, are objectively discoverable by orthodox canons of science, and the results, wherever possible expressed in numbers, are capable of being fed back into enlightened policy-making in the form of rational decision criteria applied by experts. The central problem for effective public risk management is to ensure that these decision criteria and the associated risk data are as accurate and scientifically well grounded as possible – including attempts to incorporate "the human factor", provided that such an elusive element can be reduced to terms that are tractable for engineers.

Indeed, this approach is not simply an engineer's view, for there are social science approaches to risk that fit the same conception of risk management very closely, and hence can readily be absorbed into the paradigm. The most notable case is the well established and enormously successful psychometric paradigm, which aims to identify general features of individual cognitive approaches to risk; but Mary Douglas (1994) has argued that the developing "risk amplification" and "risk communication" approaches also fit with the dominant "enlightened engineering" approach to risk, because both essentially exclude politics from the analysis.

For this dominant approach, "problems" come in the form of "irrational" behaviour by politicians and bureaucracies, unaccountably declining public trust in scientific expertise, and "maverick" social scientists who do not see risk assessment and management as a politics-free zone. Indeed, the "two chapters bad" of the Royal Society document were precisely those concerned with risk perception and management. These chapters advanced the unorthodox ideas that there might be a political dimension in the way risks are socially construed and that the fundamental doctrines of risk management are in fact inherently plural, disputable and disputed. Significantly, the document was not issued as a report of the Society (unlike its 1983 predecessor), but rather as a report of a study group, with the contents of each of the chapters attributed to its author(s).

Reasons of speed and convenience in editing were given to account for this form of publication, but even in the preface it is hinted that the need to disassociate the Royal Society from the dangerous and controversial ideas expressed in the social science chapters were the real reason for the change of format (see also Warner 1993).

A few heretics, of course, responded to the document with the opposite mantra "two chapters good, four chapters bad" – and some social scientists (notably Mary Douglas 1994) even criticized the offending social science chapters for not going far enough to embrace a politics-centred approach to risk. The "two chapters good" heretics tended to be those who are known to be critical of the dominant "enlightened engineering" approach to risk management (see Adams 1995). For example, some claim that the technocratic and quantitative emphasis in orthodox risk management tends to limit effective decision-making to a small technical elite, producing a structure that may actually blank out "safety imagination" by creating illusions of invulnerability (see Toft, this volume) and which may be highly vulnerable to "groupthink" (the well known term coined by Janis (1972) to denote unreflective adherence to unexamined assumptions). Others claim that the orthodox approach to risk management does not reflect popular attitudes to risk management (and hence lacks social legitimacy by dismissing popular concerns as "irrational"; Perrow 1984, Shrader-Frechette 1991). Even Judge Stephen Breyer (1993), in a relatively orthodox critique of conventional risk regulation politics, concludes that the solutions lie in institutional design rather than in better "risk communication" or improved technical formulae.

Our aim in producing this book is to pursue this exchange, in the belief that the rival doctrines of risk management identified in chapters 5 and 6 of the Royal Society Study Group Report (Royal Society 1992) merit some further attention. To the extent that "risk" is socially construed (involving conflicting conceptions of trust and blame), inherently involves who-gets-what distributive issues that cannot finally be solved by any simple aggregate *numéraire*, and is to some extent a "trans-scientific" area of inquiry, it follows that better understanding may be achieved through both a greater sensitivity to the rhetorical aspects of risk management debates and by a careful juxtaposition of contrasting points of view. As we explain in the first chapter, this book is designed in that spirit as a "conversation" relating to seven "what to do" aspects of risk management; and the final chapter aims to develop the "collibration"[1] approach to risk management, which was outlined in only the sketchiest form in the 1992 Royal Society document and was criticized subsequently, with some justification, for being too undeveloped to count as a serious alternative to conventional ideas of risk management.

---

1. See p. 206 for an explanation of this term.

We are very grateful to the Royal Society for granting permission for us to use much of the material contained in pages 154–67 of the 1992 report, as the basis for the introductions to the seven areas of debate. We are also indebted to the contributors who enthusiastically responded to our request to develop selected viewpoints in the risk management debate and who patiently endured the exceptionally long gestation period of the book. We also wish to thank the ESRC who funded a research seminar programme at LSE in 1991 at which some of the material contained in this book was first presented and in which the intellectual framework of the book began to develop; and to couple this with thanks to both the ESRC and LSE for their support for research to underpin the two controversial social science chapters in the 1992 Royal Society document. In addition, we have debts too numerous to mention to colleagues at LSE and elsewhere who encouraged us and helped us to develop our ideas. One of them was the late Barry Turner, who collaborated with us in writing the social science chapters of the 1992 Royal Society document and who we hoped would be a co-editor of this book. Barry Turner's ideas about how organizations create disaster helped to inspire much of the "safety" literature in British social science in the 1980s, and Barry helped us to design this book and to select appropriate contributors. We dedicate the book to his memory.

Christopher Hood   David K. C. Jones
London, December 1995

LIST OF CONTRIBUTORS

Professor David Blockley
  Department of Civil Engineering, University of Bristol
Dr Adrian Cohen
  Formerly of the UK Health and Safety Executive
Dr David Collin rid e
  formerly of Aston Business School, Aston University
Professor Sir Christopher Foster
  Coopers & Lybrand
Dr Silvio O. Funtowicz
  The Research Methods Consultancy
Professor Christopher Hood
  Department of Government, London School of Economics and Political
    Science
Dr Neil Johnston
  Aerospace Psychology Research Group, Trinity College, Dublin
Tom Horlick-Jones
  Centre for Environmental Strategy, University of Surrey
Professor David Jones
  Department of Geography, London School of Economics and Political
    Science
Professor Timothy O'Riordan
  School of Environmental Sciences, University of East Anglia
Professor Edmund Pennin -Rowsell
  School of Geography, Middlesex University
Dr Nick Pid eon
  Department of Psychology, University College of North Wales, Bangor
Dr Jerry Ravetz
  The Research Methods Consultancy
Dr Simon Shohet
  Centre for Business Strategy, London Business School
Dr Brian Toft
  Risk Analyst, Sedgwick James
Professor David Weir
  Bradford Management Centre, University of Bradford
Professor Celia Wells
  Cardiff Law School, University of Wales

CHAPTER ONE

Introduction

David Jones & Christopher Hood

Since the mid-1980s, there has been an upsurge of academic and popular interest in the subject of risk and its management. Indeed, it has been claimed that risk is emerging as a key organizing principle in social science (Beck 1992, Douglas 1992, Giddens 1990, 1991) and become "one of the most powerful concepts in modern society" (Leiss & Chociolko 1994: 3). Controversies as to how risk should be managed are now claimed to rank as "among the most bitter disagreements in contemporary society" (ibid.: xiii).

The reasons for this growth in attention are complex and debated. For some, rising risk-consciousness reflects increased expectations of health, safety and security within advanced technologically based societies, notions that have been given added status because of contemporary aspirations to sustainable development. Others emphasize the increasing numbers of people who believe they cannot control their exposure to the chance of unfair or uncompensated loss caused by the activities or decisions of others. Since the mid-1980s, large-scale high-technology developments, such as nuclear power and biotechnology, have come under continuing attack, and there have been widespread fears of adverse consequences created by unforeseen and invisible threats (such as AIDS, the BSE–CJD link, radon, electricity transmission and certain kinds of environmental pollution). Other foci of attention, blame and apprehension have been the possible adverse outcomes of human-induced global environmental change and the apparently remorseless escalation in costs of so-called "natural disasters" (reflected in the United Nations declaration of the 1990s as the International Decade for Natural Disaster Reduction). And we can add a diverse range of longer-standing concerns on issues such as resource availability, the safe operation of industrial plant, transport safety, pesticides, drugs, and the viability of certain financial institutions. Even these issues turn out to be merely the most prominent landmarks in an extensive and varied landscape, "for risk is ubiquitous and no human activity can be considered risk free" (Hood et al. 1992: 135).

Public interest and attention has stimulated the production of many books and papers on the subject of risk. But despite – or perhaps because of – this burgeoning literature, there is little sign of closure of this debate and three fundamental problems remain unresolved. First, there is no clear and commonly agreed definition of what the term "risk" actually means. Secondly, research on risk remains highly compartmentalized, fragmenting the subject into many relatively isolated subfields and specialisms (a feature referred to as "the risk archipelago" by Hood et al. 1992). Thirdly, there is continuing controversy over what-to-do issues – the "practical philosophies", principles or doctrines concerning how the different major components within the "risk environment" are best managed.

This book is intended to provide some insights into this third issue – debates over doctrines of risk management – by exploring further the competing approaches that were outlined by Hood et al. (1992) in the Royal Society Report, *Risk: analysis, perception and management*. However, before embarking on this exploration, we need briefly to examine the other two major "hot spots" in the risk literature to which we referred earlier.

The meanings of "risk" and "hazard"

It is notable that, instead of converging on agreed definitions of basic terms, the contemporary literature abounds with contradictory statements about what the words "hazard" and "risk" mean. Some authors appear to use the terms interchangeably whereas others put "hazard" at the centre of the stage, limiting risk to its original technical sense to connote likelihood or probability of occurrence (see Cutter 1993: 2). It remains to be investigated systematically how far these differences of usage can be explained by rhetorical analysis, cultural bias, the academic pedigree of the author or the nature of the problem being studied. But it can be observed that natural scientists involved with extreme events, such as floods, hurricanes and earthquakes, tend to favour "hazard", whereas social scientists and mathematicians employ "risk". This diversity of vocabulary makes the language of the subject tricky to master.

The term "hazard" generally denotes a phenomenon or circumstance perceived to be capable of causing harm or costs to human society. The anthropology of taboo testifies to the rich variety of beliefs about agents, situations or circumstances held to have the potential to cause loss, should they affect persons, property or resources by threatening life, health, emotional security, material welfare or societal institutions. Hazard is, therefore, concerned with the *cause* of perceived adverse consequence.

"Risk", by contrast, is a broader and more diffuse concept. In the usage of "normal science", it connotes the assessment of *consequence* or "exposure to the chance of loss" (Leiss & Chociolko 1994: 6). Reflecting that usage, Warner (1992: 4) argues that the term should not be restricted to the mere likelihood (probability) of an adverse impact but rather to "a combination of the probability, or frequency, of occurrence of a defined hazard and the magnitude of the consequences of the occurrence". Defined in this way, hazard becomes a component within risk, making hazard management a subset of risk management.

Such a wide range of possible hazards exist – associated with day-to-day human life, both within the natural environment and the technological, commercial, legal and sociopolitical systems – that the selection of which risks and hazards are to occupy a central place on the political agenda is a complex and much-disputed social process (Douglas 1992). There is a close analogy with the range of beliefs and interpretations of cause–effect relationships affecting the "environment" and differing views of how ecosystems work (that is, how organisms relate to one another and what the relationship is between organisms and their abiotic environment). Building on that analogy, "risk" can be said to comprise perceptions about the loss potential associated with the interrelationship among humans and between humans and their natural (physical), biological, technological, behavioural and financial environments – a complex that may conveniently be termed the "risk environment".

The risk archipelago

In view of the extremely diverse character of risk, it is no surprise that risk analysis and management is not a tightly unified and consensual field but instead consists of many distinct subdisciplines and specialisms, rather like the islands of an archipelago.

Traditionally, the field has been divided by specialists according to the type of hazard under examination: natural, technological or social (Fig. 1.1). A specialist field of natural hazard research has tended to employ human ecological perspectives to examine the "goodness of fit" between human societies and non-human physical processes (see White 1974, Burton et al. 1978, 1993, Kates 1978, Kates & Burton 1986, Bryant 1991, Smith 1992, Alexander 1993). But the term "environmental hazards" is increasingly coming to replace "natural hazards", reflecting a growing appreciation of the extent to which "natural" environmental systems are influenced by human activity to create or exacerbate hazardous conditions, for instance in desertification, floods and acidification. Technological hazards, such as explosions, collisions or systems failure, emanate from the design and management of technological systems (see Turner 1978, Perrow 1984, Lewis 1990), whereas *social hazards* emanate from

3

human behaviour alone, such as fraud, burglary, arson or terrorism (Rosenthal et al. 1989).

The apparently particular nature of each of these three different groups of hazards has led to the evolution of separate and distinct fields of study, each of which has generated its own particular literature, methodology and terminology. Further specialization has led to division into subfields consisting of groups of relatively insular hazard-specific experts. Although such conventional distinctions are understandable and may be useful, up to a point, in developing specialization, they may be becoming more of a problem than a solution to the effective analysis of some crucial aspects of risk management. The conventional distinction between "natural" hazards on the one hand, and "technological" and "social" hazards on the other (see Fig. 1.1) is precarious, because what is seen as the product of extraneous natural forces as opposed to human behaviour is culturally variable (it has to be remembered that our very notion of hazard involves impact on human wellbeing). Also, the three categories are not even analytically distinct, but, as Figure 1.1 shows, they overlap and merge to produce hybrid hazards, otherwise known as quasi-natural hazards and "na-tech" hazards. Hence, similar hazard impacts and management problems can be produced by events/accidents/incidents caused by very different trigger mechanisms, and the conventional tripartite division based on causation can obscure rather than illuminate the social processes at work, because the impact characteristics of a hazard tend to be of greater significance to those involved than its causes.

An alternative division focuses on the different levels of *scale* and *frequency* in risk management. The approach of one "island" focuses on the special problems associated with managing conspicuous high-magnitude low-probability (low frequency) events, usually termed "disasters" or "catastrophes" (see Quarantelli et al. 1986). Disaster research, which first emerged as a recognizable subfield of the social sciences in

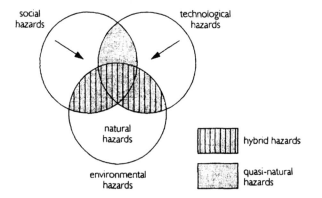

Figure 1.1 The hazard spectrum (Jones 1993).

the late 1950s, has mainly focused on the response phase of disasters, but is coming to give more attention to recovery and mitigation processes (see Drabek 1986, Comfort 1988). Other work is more concerned with incidents of lower magnitude and higher probability (such as road accidents or accidents in the home), which may, nevertheless, have a larger absolute fatality rate. Some attempts have been made to link the two islands, by constructing absolute scales for disaster using objective measurement (e.g. the Bradford Disaster Scale; see Keller et al. 1990), but such scales have been the subject of much criticism and are not widely accepted, partly because there are no simple relationships between social impact and objectively measurable aspects of scale such as death rates.

Thus, a picture emerges of an exceptionally broad, diverse and yet compartmentalized field of study. Although certain disciplines, such as quantitative risk analysis, economics and the study of government regulation, have tended to address risk management in a generic way, these approaches continue to be in a minority. Mitchell (1990) bemoans the lack of a single comprehensive integrated body of study created by the existence of three quite distinct coherent subfields: natural hazards research, disaster research and risk analysis. But the divisions by no means end there. The Royal Society's (1992) attempt to produce a multi-science approach to risk revealed a major conflict of cultures, ostensibly between social scientists stressing social construal of risk and rivalry among competing risk-management doctrines and natural scientists with a more objective view of risk and a more uniform vision of good practice. But the "four chapters good, two chapters bad" interpretation of the problem (see Preface) obscures major differences *within* the social and natural sciences, among practitioners and even between cultures. According to Turner (1994: 148) "we are now in a situation where no single view of risk can claim authority or is wholly acceptable", whereas Beck (1992: 29) goes one step further by claiming "There is no expert on risk".

This book is an attempt to go beyond the over-simple "four chapters good, two chapters bad" view of the risk management world and to build some further bridges between the islands of the risk archipelago. In the following chapters, hazards of different type, scale and frequency are discussed and most of the major disciplinary "voices" in risk management are included, with contributions to the debate from perspectives based in law, engineering, economics, psychology, sociology, political science and geography. To some extent, this approach represents a search for a more *generic* approach to risk (an issue that is discussed further in Ch. 9). But it also reflects a view that much *is to be learned about risk from an* approach that does not divide the world too neatly into separate specialisms, and to avoid a situation in which the inhabitants of each island re-invent the wheel for themselves. A wide-ranging approach to risk seems appropriate to an attempt to bring out the major voices in the debate and identify the areas of discordance and harmony.

So, what is risk management?

What does it mean to "manage" risk, given the variety and dimensionality of the phenomenon? It is certainly easier to say what risk management *is supposed to do* – at least in an abstract and formal way – than what it ought to be *for*. Like any other form of management, risk management can be understood as a process involving the three basic elements of any control system (cf. Dunsire 1978: 59–60) namely:

- goal-setting (whether explicit or implicit)
- information gathering and interpretation
- action to influence human behaviour, modify physical structures or both

Each of these three elements is linked to fundamental questions about risk management that are easier to ask than answer (cf. Shrader-Frechette 1991). Who is to bear what level of risk, who is to benefit from risk-taking, who is to decide, and who is to pay? Where is the line to be drawn between risks to be managed by the state, and those to be managed by individuals, social groups or corporations? What information is needed for "rational" risk management and how should it be analyzed using what "justice model"? What actions make what difference to risk outcomes? Who evaluates success or failure in risk management and how? Who decides on what should be the desired trade-off between different risks? There is no general consensus on such questions, yet life-or-death risk management does – and must – go on in all societies through some sort of institutional process (Douglas 1987).

The term "risk management" means different things to different people, depending partly on which of the islands in the archipelago they inhabit. Some adopt a rather restricted technocentric view, building on the technological hazards literature and putting the emphasis on the safe operation of hazardous processes, technological systems and engineered structures. In a business context, the term means financial provision for risks. In politics, it is sometimes used to refer to the handling of issues that may threaten a government's electoral fortunes (see New Zealand State Services Commission 1991: 53–60). In public policy, "risk management" has been commonly used to refer to an analytical technique for quantifying the estimated risks of a course of action and evaluating those risks against likely benefits. In this book, we have decided to adopt the broadest view in an attempt to encompass all of the major senses of risk management, sometimes at the expense of precision. Accordingly, for

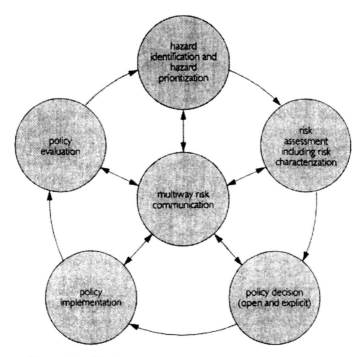

Figure 1.2 The risk management cycle (after Soby et al. 1993).

these purposes, risk management means all regulatory measures (in both public policy and corporate practice) intended to shape the development of and response to risks. Risk management in this sense means a range of related activities for coping with risk, including how risks are identified and assessed and how social interventions to deal with risk are monitored and evaluated. Figure 1.2 depicts a set of risk management processes. Formal doctrines of risk management are usually based on two presuppositions: that risk is acceptable only if it is outweighed by demonstrably greater aggregate benefits and (as with total quality management doctrines) that there has to be a continuous striving to reduce the level of risk to a point where it is held to be "tolerable", "as low as reasonably achievable", or to meet some other criterion of social acceptability. But doctrines of that type rarely refer explicitly to distributional questional questions (as opposed to aggregated social costs and benefits) or to whose view of what is "tolerable" or "reasonable" is to be taken as the dominant one.

Some have argued (National Research Council 1983) that risk *assessment* and risk *management* are overlapping, but separate, tasks. They claim that the former is mainly scientific and concerned with the establishment of probabilities, whereas the latter is primarily legal, political and administrative. But this tidy distinction between "scientific" assessment and "political" or "bureaucratic" management has been contested by those who argue that it is impossible to disentangle social values and world-views from the process of identifying, estimating and evaluating risks (Douglas 1985, Rayner & Cantor 1987; Zimmermann 1990: 9–10; Freudenberg 1992), and that, at least from a social viewpoint, it is unhelpful to conceive of risk as if it were a single uniform substance, such as "phlogiston" (Watson 1981). The existence of such a "substantial literature documenting the strong influence of the social and political context of risk assessment" (Kraus et al. 1992: 230) has led to the rejection of this claimed distinction (Leiss & Chociolko 1994), because of the widespread recognition that qualitative judgements, often heavily contested, play a major role in *both* the assessment *and* management of risk.

Dimensions of the risk management debate

Central to contemporary politics and public management is the question as to what are the principles that should govern the identification, measurement and regulation of risk. If there were agreement on an answer, then we might be led to think that the world would be a safer place to live in.

Unfortunately, the number of answers offered to this question is exceeded only by the further questions to which the answers give rise. Current debates about risk management reflect competing world-views. For those engaged with practical risk management or those involved in policy development in the area, it may be difficult to understand the significance of the disputes about different options in risk management, especially since these views are often expressed forcefully and emotively. It is helpful to try to identify elements that keep cropping up in these debates, and for this reason Table 1.1 sets out seven commonly recurrent sets of opposing views about how risk management should be handled.

The remainder of this book develops this schema, focusing on these seven issues of contention, which appear to run through many of the diverse areas of risk management. The seven differences in emphasis are used as opposed positions in debate. The chapters that follow are intended to illustrate, directly or indirectly, some of the permutations these seven positions may take, while addressing practical issues that are central to the problem of coping with risk.

Exploring these sets of controversies is a simple way of mapping many of the contours of the risk management debate. However, like all maps, the scheme presented here is a simplification of reality. For the sake of clarity, the positions presented have been grouped in opposing pairs, as mirror images of one another; or, to look at it another way, as ideal types.

In reality, few people would adopt the extreme variant, and a paler version of each position, often held in combination with others, is more common. Some of the positions discussed are unconventional, but to identify the poles of a debate is not to imply that each pole is of equal weight (in fact more often the two sides represent orthodoxies and heresies) and to identify a position is not necessarily to endorse it.

In short, what the reader should be able to gain from this book is a sense of the areas of debate and to the range of competing doctrines and beliefs

Table 1.1   The seven most prominent recurrent sets of opposing views in risk management (Hood et al. 1992).

Doctrine	Justification	Counter-doctrine	Justification
Anticipationism	Apply causal knowledge of system failure to ex ante actions for better risk management	Resilience	Complex system failures are not predictable in advance and anticipationism makes things worse
Absolution	A "no-fault" approach to blame avoids distortion of information and helps learning	Blamism	Targeted blame gives strong incentives for taking care on the part of key decision-makers
Quantificationism	Quantification promotes understanding and rationality, also exposes special pleading	Qualitativism	Proper weight needs to be given to the inherently unquantifiable factors in risk management
Design	Apply the accumulated knowledge available for institutional design	Design agnosticism	There is no secure knowledge base or real market for institutional design
Complementarism	Safety and other goals go hand in hand under good management	Trade-offism	Safety must be explicitly traded off against other goals
Narrow participation	Discussion is most effective when confined to expert participants	Broad participation	Broader discussion better tests assumptions and avoids errors
Outcome specification	The regulatory process should concentrate on specifying structures or products	Process specification	The regulatory process should concentrate on specifying institutional processes

impinging on a field that cannot be reduced to scientific certainties. Risk is inherently bound up with uncertainty. If this book is an attempt at a map, then it is the sort of map likely to contain uncharted regions, where there may be dragons.

CHAPTER TWO

Anticipation
in risk management:
a stitch in time?

ANTICIPATIONISM VERSUS RESILIENCE

Risk management in practice typically involves some mixture of anticipation – "looking forwards", and resilience – "bouncing back". One key element of the risk management debate turns on where the emphasis should be laid between the two.

"Anticipationists" argue for extra weight to be given to measures designed to detect in advance the clues that signal potential threat in physical or organizational structures, and to act on those clues, even before scientific proof "beyond a reasonable doubt" has been obtained. Such an approach means laying more emphasis on methods of *ex ante* detection and prevention, and on regular "health checks" or "audits" of potentially dangerous organizations, locations or structures.

The case for adopting a more anticipationist or proactive approach is made in various ways. Some argue that the increasing complexity and uncertainty of contemporary society require an extension of precautionary "just in case" regulation, particularly in the field of pollution control. For example, Tait & Levidow (1992) note: "The *Versorgensprinzip*, or precautionary principle, originally enunciated by the West German government in 1976, has gradually become a focus for creative thinking on these subjects throughout the EEC and more widely. The precautionary principle is proactive in that it advocates the implementation of controls of pollution without waiting for scientific evidence of damage caused by the pollutant(s), and without necessarily requiring consideration of the relative costs and benefits of regulation to industry or the public".

A good example of the precautionary doctrine in operation relates to the perceived safe levels of nitrate in drinking water. Concern about a possible link between nitrate levels and stomach cancers, as well as "blue baby" syndrome (infantile methaemoglobinaemia), led to the EEC formulating directives in the 1980s setting down strict limitations designed to control the problem. These precautionary measures will take some time to implement and will involve very substantial costs in some regions, even though the harmful effects of nitrate levels twice those set down by the EU have not been confirmed by subsequent scientific research. Clearly, the issue of exactly how far the "precautionary principle" ought to be taken, and how far public policy should run ahead of clear scientific findings, cannot be resolved according to any authoritative technical formula (see O'Riordan & Cameron 1994).

The dilemma is particularly controversial when it relates to the trial of new drugs, which may, on the one hand, produce harmful side-effects in humans, but may also have life-saving benefits. In such cases, different types of risk must unavoidably be traded off against one another (cf. Shrader-Frechette 1991: 67) and which course of action will best satisfy a "least harm" principle may be impossible to establish by purely "decisionist" methods (cf. Majonre 1989: 12–20).

A very different argument for a more "anticipationist" stance goes that, in hindsight, disasters can often be interpreted as "waiting to happen" and may even be linked to particular types of *organizational* culture and structure. Those who adopt this view point to research on systems failure, carried out in several countries, showing that major incidents are typically produced by a concatenation of smaller slips, errors and malpractices, which themselves could have been reduced, if not eliminated, by improved everyday work practices, better training, better "safety culture" and more frequent and more adequate safety audits (cf. Baldissera 1987, Pidgeon 1988, Turner 1978, 1991, HSE 1990a). Failure to learn effectively from previous errors is often claimed to be at the heart of many disasters (see Handmer 1992: 117). Horlick-Jones et al. (1991) take the concept of a disaster-producing "system" even further, by examining organizations within their environment, which includes such factors as resource allocation, task overload and regulation. It is the complex interaction of organizations and their environment that creates vulnerability.

This second strain of anticipationism bases its claims on information about the chains of causation leading up to both rapid-onset and slow-onset disasters. The argument is that particular organizational patterns, such as regulatory conflicts of interest, secrecy and lack of broad participation, are frequently associated with major failures. Research designed to identify such associations can be fed back into risk management practice through an extended process of "hazard audits" (Toft 1990). Broadly similar arguments can be found in debates about the management of geophysical hazards (earthquakes, hurricanes, floods), where there is a strong body of opinion that various types of anticipatory planning are essential to risk reduction (Foster 1980, Wijkman & Timberlake 1984, Smith 1992, Burton et al. 1993). However, in these cases it has to be noted that technocentrism plays a crucial role, for much of the anticipatory activity is focused on monitoring systems, warning systems and the development of engineered defensive structures, as is clearly manifest in the documentation for the UN International Decade for Natural Disaster Reduction (Lechat 1990, UNDRO 1990, Smith 1995).

Although anticipationism may appear intuitively sensible, there are many writers and practitioners who advocate a much more cautious "wait and see" approach. Institutional analysts such as Wildavsky (1985, 1988) have argued that disasters and system failures often look "predictable" only with the benefit of hindsight, and that many such cases involve high-dimensional dynamic systems whose behaviour is inherently difficult or even impossible to forecast, because of their complexity. The implication is that risk management regimes should be designed to promote resilience against unexpected catastrophes, rather than to rely on being able to spot them coming in time to take action to prevent their occurrence or lessen their impact. It can even be claimed that placing too much emphasis on the anticipation of adverse outcomes may actually *contribute* to the crises it seeks to avoid, by a *"Titanic* effect" (or the related "levée syndrome" in natural hazard management), which reduces the capacity to respond to the unexpected and increases the shock when things actually do go wrong. For the advocates of resilience, the emphasis of a risk management regime should be directed more on promoting the capacity to cope with the unexpected, for example by relief, emergency action, rescue, insurance, and so on (Cuny 1983).

Most of the anticipation–resilience debate is about what should be done at the *margin*, not about absolutes. Few serious contributors to the debate would put all the emphasis on one approach rather than the other, and it is a normal principle of sound design to incorporate both the lessons of previous failures and forethought about likely future ones. Indeed practitioners use devices of piloting and monitoring (e.g. in post-occupancy evaluation of building use) as a means of bridging anticipationism and resilience.

The anticipation–resilience controversy has many possible shades and variants. In the case of increasingly complex technological systems, the issue comes down to a discussion of the extent to which system failures (particularly those involving human organization) are low- or high-dimensional dynamic systems, or more precisely what kinds of failure involve the one and what involve the other. In principle that issue might be clarified by more systematic investigation, but it seems likely that it is in large part a "trans-scientific" issue: that is, a problem that in theory is investigable by orthodox scientific methods but in practice cannot be so investigated because of limits on time, resources or morally sanctioned behaviour (Weinberg 1972), and hence must, unavoidably, be debated through a process of rhetoric.

Some of these themes are explored further in the three sections that follow. First, David Jones examines the growth in significance of anticipationism in the context of managing risks arising from natural hazards.

He argues that the current phase of "hazard management", which focuses on the deployment of science, technology and engineering, is likely to be followed by a second phase focusing on risk management through the reduction of human vulnerability. Then David Blockley writes on hazard engineering from the perspective that one can manage hazard but only anticipate risk (*pers comm.*). He argues that the two "world views" of hazard, risk and safety – the one technical/engineering, the other human and organizational (social science/management) – need to be unified through a new discipline of hazard engineering based on Turner's (1978) model of accident incubation. The two basic activities of hazard engineering are hazard auditing and hazard management, and success would be dependent on the adoption of five key points:
- the explicit recognition of hazard as a sociotechnical problem
- use of the systems approach
- recognition that reflective practice is preferable to technical rationality
- acceptance that responsibility is preferable to reliability
- acceptance that the limits of models should be articulated.

Finally, David Collingridge argues that, in the case of technological development, the future is often so uncertain that resilience should be the favoured option. This resilience is best achieved, he believes, through the adoption of flexibility and diversity.

ANTICIPATING THE RISKS
POSED BY NATURAL PERILS

David K. C. Jones

Introduction

It is widely appreciated that uncertainty is the enemy of anticipation. Well known statements, such as "While limited to only one past and one present, every society faces a multiplicity of potential futures" (Sewell & Foster 1976), appear to suggest that a "wait and see" policy is preferable to one that attempts to anticipate the possible adverse consequences of future events, especially those arising from the workings of complex environmental systems largely outside human control. But this is far from the case, for there are many "anticipationist" natural scientists who are increasingly optimistic regarding the evolving ability of human society to cope with cost-inducing physical environmental phenomena, popularly known as "natural hazards" or "natural perils", but better termed "environmental hazards" (see Ch. 1). They can point to an impressive and ever-growing range of examples where anticipatory measures have saved lives, reduced suffering, limited damage and destruction, and restricted adverse economic consequences, thereby indicating that more can be done to bring hazard losses (the so-called "natural tax" or "natural rent") down to tolerable levels. This contribution is written from their perspective, to show the scope for anticipationism.

Changing views of natural perils

It has long been recognized that the physical and biological components of the geosystem include elements that are detrimental to human aspirations. Initially, this knowledge was fragmentary and rudimentary, but it has become deeper and better integrated as a consequence of scientific enquiry. Just as it took time to establish the actual causes of diseases, such as malaria or cholera, the same is true of geophysical phenomena such as riverine flooding, tropical revolving storms, and earthquakes. Slowly at first, but then with increasing speed, scientific research, bolstered by growing technological capability, has explained the great majority of diverse unusual phenomena that were previously considered to be of supernatural origin and thus termed "Acts of God". As a consequence, rapid onset events of spectacular appearance and great violence, such as tsunamis (giant sea-waves), volcanic eruptions, earthquakes, tornadoes (and water spouts) and other "unusual" meteorological phenomena, have become "internalized" within knowledge and have come to be seen as normal, rather than abnormal, features of the ecosphere. It follows, therefore, that human views of hazard impacts have also had to change, so that they are increasingly coming to be seen as part of "normal life" rather than externally imposed disruptions (Hewitt 1983). This leads naturally to the view that they are capable of analysis and assessment, with the objective of identifying strategies that can be adopted in order to minimize their future impact on society. As a consequence, recently published books on "natural hazards" have moved away from being anecdotal, with an emphasis on catalogues of mayhem and descriptions of sensational catastrophic (extreme) events, and instead have increasingly focused on the more optimistic theme of hazard loss reduction through anticipatory actions based on scientific assessments of problems (Palm 1990, Bryant 1991, Kreimer & Munasinghe 1991, McCall et al. 1992, Smith 1992, Alexander 1993, Burton et al. 1993 – merely to mention the most accessible texts).

Anticipating natural perils

Natural scientists will argue that this growth in knowledge has allowed a significant shift in emphasis regarding human interaction with the physical environment, in that the traditional focus on post-impact recovery is being increasingly replaced by more forward-looking hazard loss reduction practices. They claim that, once the nature of physical phenomena are understood, then it is possible to use mapping, monitoring and modelling to inform anticipatory measures in the following ways:

- Monitoring of occurrence, together with surveys of past occurrences, reveal the *spatial distribution* of events thereby facilitating the delimitation of *hazard zones* (i.e. areas within which specific phenomena are known to occur) and their division into zones of differing orders of threat, as determined by combinations of magnitude and frequency of occurrence *(hazard zonation)*, a process that may be continued to high levels of detail *(microzonation)* if the data allow (Fig. 2.1). Many examples of zonation maps exist covering a range of phenomena (see Foster 1980), including riverine flooding (Kates & White 1961), earthquakes (Degg 1992a, Degg & Doornkamp 1989, 1990, 1994), slope instability (Jones 1992, 1995) and tornadoes (Wolford 1960, Fujita 1987).
- Records of occurrence reveal magnitude–frequency relationships that may be global, regional or local, depending on levels of information

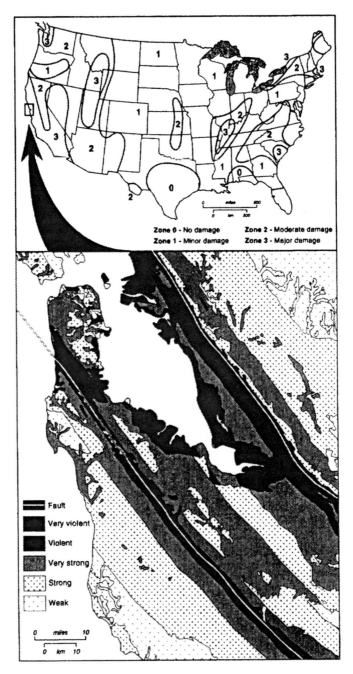

Figure 2.1 Zonation and microzonation maps of earthquake hazard, USA and part of San Francisco area, California. The broad national generalizations conceal major local variations as detailed in the microzonation map developed for maximum expected ground shaking. The microzonation map provides a clear framework for a broad range of anticipatory actions, including planning controls, site investigation procedures, building codes, emergency planning and insurance premiums (After Jones 1991, adapted from maps produced by the US Office of Emergency Preparedness and the US Geological Survey).

and need. Such magnitude–frequency distributions usually reveal a spectrum of occurrence with many small events (in terms of volume or available energy), referred to as *low magnitude–high frequency events*, and few large events, known as *high magnitude–low frequency events*.

• Monitoring of environmental conditions, together with modelling, allows the identification of circumstances suitable for creation of hazardous events and the development of hazard generation scenarios.

This knowledge provides the basis for the three types of *prognostication* that underpin anticipatory actions:

• the ability to *predict* hazard potential within an area or at a site, by subjecting records of past occurrence to various types of *extreme value analysis* in order to yield quantitative assessments of likelihood and frequency of occurrence of differing magnitudes of hazard events (expressed as recurrence intervals/return periods, frequencies or probabilities)

• the ability to monitor environmental conditions in order to foretell the occurrence of specific events in terms of location, timing and magnitude (*forecasting*), so as to allow warnings to be issued and emergency actions to be undertaken

• the ability to estimate the likely consequences of impact through the use of various predictive devices, including scenario-building using analogues (*futurology*), which allows exploration of the variable linkage between hazard magnitude and impact intensity.

These three distinct components of *prognostication*, of which *prediction* is by far the most important because it underpins the other two, provide the basis to anticipate threat, the realization of threat and the requirements for sustainable recovery. Workers in the field of natural hazard research have for long recognized five main groupings of anticipatory measures (or adjustments) that can be undertaken to lessen impacts and reduce potential losses (see Burton et al. 1978):

(a) Actions designed to limit the magnitude and/or frequency of potentially threatening circumstances by modifying environmental conditions; examples include cloud seeding (to reduce hurricane ferocity, disrupt the formation of large hailstones, suppress lightning, clear cold fogs and increase precipitation), land-use management (to reduce flooding and slope instability), retaining structures (to limit avalanching and slope instability), emission controls (to reduce smogs), access controls (to limit avalanching) and the controlled stimulation of events in order to limit their impact potential (for example, the use of explosives to trigger avalanches and slope failures, or the possible future use of explosives and/or pressure-injected liquids to release crustal stresses by means of small earthquakes, rather than let them build up to generate potentially destructive major shocks).

17

(b) Actions designed to constrain/control threatening events by means of deflection (tree belts, control structures), the dissipation of available energy (wave spoilers, the use of dams to reduce flood magnitudes by absorbing some of the flood wave (flood routing)), or defensive structures designed to raise impact threshold (barriers, walls, levées).

(c) Actions designed to modify loss potential through improved prognostication and risk communication, so that the risk to people, activities and structures is minimized. Actual measures include using hazard *prediction* to inform land-use zonation policies, planning controls and building codes/ordinances; using *prediction and futurology* to establish proper emergency action plans and procedures; and using *forecasting and risk communication* (including warning systems) to ensure that requisite actions are taken when the threat materializes (evacuation, emergency response, rescue, relief).

(d) Actions taken to plan for financial losses in terms of insurance schemes, development of reserve funds, and so on.

(e) Actions taken to spread potential losses through reinsurance, governmental responsibility, international relief, and so on.

The twentieth century has witnessed considerable advances in all five groups of adjustment, but most conspicuously in groups (a) and (b), which can be termed *hazard management*, as well as group (c) which, together with the remaining two groups, can be referred to as *vulnerability management*. Groups (a–c) are all heavily technocentric in character, involving the application of science, technology and engineering, and support the widely disseminated view that technocratic solutions are the best way of minimizing the risks posed by natural perils.

Evidence of the success of such approaches can be drawn from many sources, especially in the case of earthquakes, where the similarly sized 1988 Armenian and 1989 San Francisco shocks (Richter magnitudes 6.9 and 7.1 respectively) resulted in very different levels of destruction (US$14 billion and US$7 billion) and markedly contrasting death tolls (25 000 and 63 respectively). Just as impressive is the decline in lives lost in the continental USA because of hurricanes over the past century (Fig. 2.2), which reflects more resistant buildings, greatly improved forecasting and accuracy of warnings resulting from the use of aircraft and satellites, high levels of public awareness and well tested evacuation procedures.

Some ardent technocentrics will go even further. To them, the catalogues of mayhem, depressing statistics of casualty rates by hazard type or by geographic region, and the graphs of remorselessly increasing numbers of impacts, death tolls or estimated losses, which inevitably appear in the early sections of books on "natural hazards", are the cause for optimism, rather than pessimism. They argue that the graphs of increasing losses (Fig. 2.3) and numbers of impacts (Fig. 2.4) are a natural consequence of

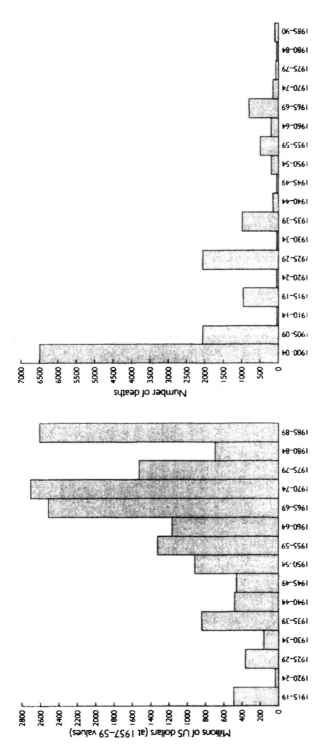

Figure 2.2 Damage and deaths from hurricanes in the continental USA, 1900–1990. Even if the 6000 deaths from the September 1900 Galveston hurricane are excluded, the trend of deaths is still downwards, despite population growth.

better information and increased exposure to loss arising from population growth, urbanization, wealth creation, economic restructuring and infrastructure development, and would be even greater but for anticipatory actions. Certainly, there is some evidence to indicate that an increasing proportion of the population is surviving major hazards such as tropical revolving storms and earthquakes (Fig. 2.5) and there is no evidence that the rate of "natural tax" is increasing.

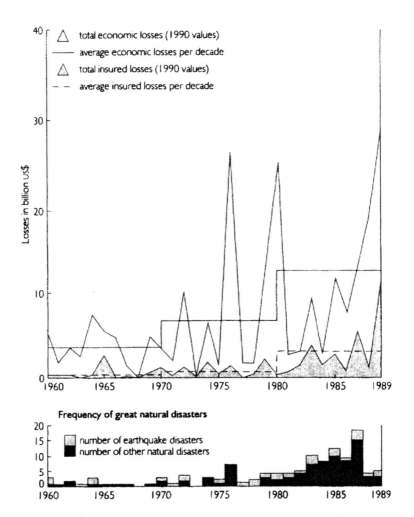

Figure 2.3 Losses from natural disasters 1960–1989 (*source:* Munich Reinsurance).

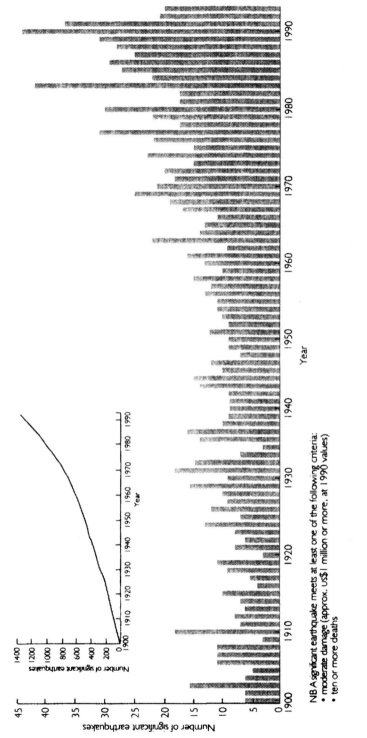

Figure 2.4 Increasing frequency of earthquake impacts shown per year (main diagram) and as a cumulative curve (inset). The growth in recorded occurrence since the early 1960s reflects better data gathering and increased vulnerability, rather than rising magnitude and frequency of significant earthquakes. (*Source:* Catalog of Significant Earthquakes, NGDC 1994).

NB A significant earthquake meets at least one of the following criteria:
• moderate damage (approx. US$1 million or more, at 1990 values)
• ten or more deaths

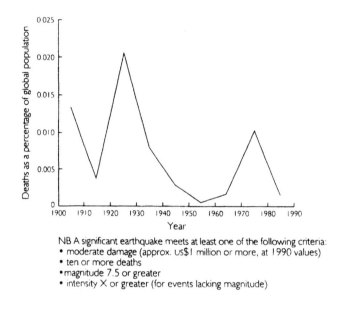

NB A significant earthquake meets at least one of the following criteria:
* moderate damage (approx. US$1 million or more, at 1990 values)
* ten or more deaths
* magnitude 7.5 or greater
* intensity X or greater (for events lacking magnitude)

Figure 2.5 Deaths per decade in significant earthquakes as a proportion of average global population per decade, 1900–89 (*Source:* Catalog of Significant Earthquakes, NGDC 1994).

The International Decade for Natural Disaster Reduction

This technocentric optimism is clearly displayed in the anticipatory philosophy promulgated by the United Nations when declaring the 1990s the International Decade for Natural Disaster Reduction (IDNDR). The unanimous passage of General Assembly Resolution 42/169 on 11 December 1987 was based on the belief that more could be done to limit the impact of "natural" hazards, which were considered to be continuing to pose unacceptably high and escalating costs to human society through disastrous impacts estimated annually to claim an average of 150000 lives, make 40 million people homeless and result in immediate damages measured in tens of billions of dollars. Resolution 42/169 (1987) referred to "a decade in which the international community, under the auspices of the United Nations, will pay special attention to fostering international co-operation in the field of natural disaster reduction". In view of the apparent disproportionate effect of so-called "natural disasters" on developing countries with over 90 per cent of disaster casualties worldwide, it is unsurprising to note that the objectives of IDNDR were subsequently refocused towards the needs of the developing world (Resolutions 43/202

(1988) and 44/236 (1989)), as illustrated by the statement that the purpose of the IDNDR is to:

reduce through concerted international actions, especially in developing countries, loss of life, property damage and social and economic disruption caused by natural disasters such as earthquakes, windstorms (cyclones, hurricanes, tornadoes, typhoons), tsunamis, floods, landslides, volcanic eruptions, wildfires, grasshopper and locust infestations, drought and desertification and other calamities of natural origin

by the achievement of the following five goals:

- to improve the capacity of each country to mitigate the effects of natural disasters expeditiously and effectively, paying special attention to assisting developing countries in the assessment of disaster damage potential and in the establishment of early warning systems and disaster-resistant structures when and where needed:
- to devise appropriate guidelines and strategies for applying existing scientific and technical knowledge, taking into account the cultural and economic diversity among nations;
- to foster scientific and engineering endeavours aimed at closing critical gaps in knowledge in order to reduce loss of life and property;
- to disseminate existing and new technical information related to measures for the assessment, prediction and mitigation of natural disasters; and
- to develop measures for the assessment, prediction, prevention and mitigation of natural disasters through programmes of technical assistance and technology transfer, demonstration projects and education and training tailored to specific disasters and locations and to evaluate the effectiveness of those programmes.

The fact that the Decade has failed to capture the attention of the public or the media has little to do with the laudability, desirability or practicality of the avowed aims. The central problems have lain in the lack of available finance, in the onus placed on individual nation states to determine their own programmes, which has inevitably resulted in bureaucratic appropriation and the uncoordinated fragmentation of effort, and in the accident of timing.

"Natural hazard" impacts attained prominence in the late 1960s and 1970s after three decades of relative quiescence, coincident with dramatic developments in the media. Thus, the 1970 Bangladesh floods (over

200 000 killed), the 1976 Tangshen (China) earthquake (242 000 killed) and the growing realization of the true scale of human suffering caused by desertification in the Sahel, resulted in a dramatic increase in awareness of "natural" hazards. Inevitably, anxiety about the ability of some societies to cope with environmental hazards resulted in calls for a coordinated international response, and the notion of a decade was not long delayed (an International Decade for Natural Hazard Reduction was first proposed by Dr Frank Press in his Presidential Address to the 8th World Conference on Earthquake Engineering in 1984). However, by the time the IDNDR had become a reality, international attention had been diverted to other issues: the collapse of communism and the fragmentation of the Soviet Union, world recession, the disintegration of the former Yugoslavia, the Gulf War and its aftermath, terrorist activity, and so on. In addition, current high levels of concern over the problematic outcomes of the cumulative, diffuse and insidious "elusive" hazards (Kates 1985) of global warming and stratospheric ozone depletion, have further assisted in diverting attention away from the traditional high-energy rapid-onset geophysical events that are the primary focus of IDNDR. Indeed, there has been some questioning as to whether such "natural" hazards are really of sufficient international significance as to require a "Decade", and whether it would not have been better to focus attention on specific problematic hazards and identified hazard-prone locations or economies. Although there is little doubt that much good will emerge from IDNDR, including, it is hoped, vastly improved data on the magnitude and frequency of losses and better evaluations of costs and significance of impacts at global, regional and national scales, it is disappointing to note that the May 1994 World Conference on National Disaster Reduction (held in Yokohama, Japan) failed to arouse more than minimal media attention (see Douglas 1995, for details of the UK contribution).

But there is an even more fundamental criticism of IDNDR, which focuses on what is known as the "dominant view" of natural hazards (Hewitt 1983). This traditional view grew out of the development of natural hazards research in the USA in the mid-twentieth century, where a human-ecological approach was adopted to examine the apparent mismatch between the "human use system" and the "natural events system". The result is the "behavioural paradigm", which envisages "natural" hazard impacts as the consequence of the lack of adjustment between human societies and the physical environment: a mismatch that is best minimized by focusing attention on the cause of losses (i.e. the physical environment) and, through the use of science and technology, limiting impacts on society by attempting to control phenomena and by the creation of engineered defences, the construction of durable structures, the development of warning systems and other methods of public protection.

The universal value of this technocentric approach came to be disputed over a decade ago by authors who pointed out that hazard impacts (i.e. disasters) were not solely attributable to physical phenomena but that humans contributed significantly, not only as individuals (the focus of environmental perception studies) but also as human societies (O'Keefe et al. 1976, Hewitt 1983, Susman et al. 1983, Wijkman & Timberlake 1984). As a consequence, the emphasis has shifted so as to interpret hazard impacts as the consequence of the interaction between physical phenomena and the vulnerable facets of society (Fig. 2.6), where vulnerability can be defined as "the characteristics of a person or group in terms of their capacity to anticipate, cope with, resist and recover from the impact of a natural hazard" (Blaikie et al. 1994, 9). As vulnerability is determined to a large extent by socioeconomic and political factors, the awareness of social structures as being a significant contributor to hazard losses resulted in the establishment of the "structural paradigm". Originally this approach was focused on local and national structures in a developing world context, but it has come to be expanded in recent years, following the recognition that international factors often play an important role in determining local vulnerability, thereby suggesting that "political economy paradigm" could prove a better name. These views are best presented in the recent book by Blaikie et al. (1994) which states "analysing disasters allows us to show why they should not be segregated from everyday living, and to show how the risks involved in disasters must be connected with the vulnerability created for many people through their normal existence", and results in the general conclusion that "the social, political and economic environment is as much the cause of disasters as the natural environment".

The fundamental failing of IDNDR is that it continues the tradition of viewing "natural" hazard impacts as "special", "unusual" and "carefully roped-off from the rest of man [sic] environmental relations" (Hewitt 1983). This, together with the strong "technological fix" approach, has the effect of marginalizing usefulness. Only as the IDNDR has proceeded has "education" come to be recognized as an important element in hazard loss reduction and disappointingly few signs exist that consideration of the social science dimensions of the problem will ever figure prominently in the programme. This neglect is unfortunate, for only when consideration of hazard becomes part of normal planning processes will any real advances be made in terms of further reducing disaster losses.

Problems for progress

The above remarks regarding IDNDR are not intended as criticisms of anticipation but rather to show that the main thrust of the UN programme

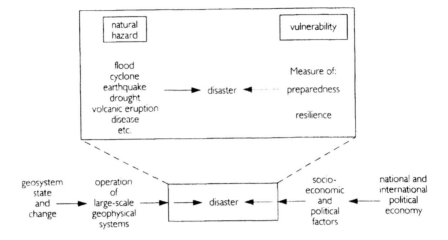

Figure 2.6 Hazard impacts result from the interaction of hazard agents with the vulnerable aspects of society. The traditional behavioural paradigm focuses on the physical causes of hazard impacts (left-hand side); the more recent structural paradigm emphasizes the role of society in generating losses (right-hand side) (after Jones 1993).

is tackling only one part of the problem. Human–environment interactions are of such complexity, involving the interplay of complex environmental systems subject to change and "human use" systems that are also dynamic, that focusing on the containment of hazard through the application of science, technology and engineering simply represents a pragmatic quest for a partial solution. However, if the anticipatory measures advocated within IDNDR are seen as merely providing an essential first step to be built upon subsequently, then the world will indeed come to be a much safer place.

But even the anticipatory approaches currently favoured for natural perils management are not without their fundamental problems. The emphasis is still predominantly on "hazard management", yet knowledge of hazardous events is still partial, the influence of changing conditions through time remains uncertain, and attempts to apply knowledge are often only partially successful because of problems of implementation. The following discussion briefly summarizes some of the difficulties.

Quality of records

Despite dramatic advances in recent decades, predictive ability remains limited, because the quality of records is spatially very variable and for many areas is best described as fragmentary or non-existent. Even where records do exist, time series may be too short or incomplete, and quality of past observations or analyses may be causes for concern. This raises problems with assessments of the geographical distribution of hazardous events, the true magnitude and frequency of potentially highly destructive extreme (catastrophic) events, the magnitude–frequency characteristics of specific hazards at particular locations (areas or sites) and, therefore, the hazardousness of place. The solution lies in long-term investment in mapping and monitoring of hazardous processes and events: a goal that may well be brought nearer as a direct consequence of IDNDR.

Temporal chan es

Predictive ability also suffers from uncertainty regarding the significance of temporal changes. Frequency–magnitude characteristics of phenomena are profoundly affected by changes to environmental conditions because of the complex interaction of anthropogenic, geodynamic and extraterrestrial factors. These influences for change differ in scale, intensity, direction and duration, and produce variations from the local to the global that are irregular (e.g. climatic changes as a result of the El Niño effect or volcanic eruptions) or cyclical (e.g. climatic changes as a result of sunspot cycles or the Croll–Milankovitch mechanism). As a consequence, predictive data based on observations of the recent past may prove unrepresentative of the future, especially in the case of climatic and riverine hazards. Human impact is also of significance at all scales up to the global, and considerable uncertainty exists as to how human influences will interact with natural mechanisms of change in the future so as to limit predictive ability further.

Global environmental chan e

Great uncertainty surrounds the potential consequences of global environmental change (GEC), and most especially of global warming, where change has been equated with threat. Many "Doomsday" scenarios have been advanced hypothesizing widespread inundation of coastal areas as a result of dramatic rises in sea level; major rapid shifts in climatic belts resulting in significant disruptions to ecosystems and accelerated extinction rates of plant and animal species, increased magnitude, frequency and extent of climatic related hazards (including droughts, floods and the formation of super-hurricanes or hypercanes), disruption to agriculturally based economies and the creation of huge numbers of displaced persons (referred to as "environmental refugees") inevitably resulting in strife. Although the scientific basis of global warming continues to be debated, the best currently available comprehensive assessments to AD 2100 suggest limited impacts that are likely to be far less significant than those generated by human population growth over the same period. However, there is still growing concern that an unconstrained greenhouse-gas forcing mechanism may result in non-linear responses that become rapid, possibly sudden, at some point in the future. This concern has inevitably led to disagreements between anticipationists as to the most desirable response. Some consider that monitoring and modelling will provide scientific information in time to guide anticipatory decision-making; others argue that the potential consequences are so substantial as to make urgent precautions essential, including control of causative agents.

Predictive information

The utility of predictive information is also a cause for concern, because increasingly sophisticated assessments of hazard have to be superimposed upon long-established patterns of activity and decision-making processes. Zoning and microzoning policies may appear highly desirable, but are often difficult to implement in practice, because of adverse effects on land and property values and previously formulated plans. Similarly, progressive changes to building codes/ordinances in the light of improved knowledge often encounter resistance because of the inevitable increased costs of construction or required expenditure on retrofitting. Even in the case of the insurance industry, where the availability of improved information on the spatial and temporal dimensions of hazard and risk would appear to be of crucial importance, the reality can be very different and can range from increased financial insecurity, because of adverse selection, to withdrawal of cover from the most hazard-prone locations (Palm et al. 1990, *The Economist* 1994). One especially ominous development is that, although general reluctance to invest heavily in hazard prediction continues, there is a growing willingness to contemplate litigation, should unanticipated impacts occur, so that the phrase "a little knowledge is a dangerous thing" can take on a new meaning. Predictions of hazard are, therefore, fraught with difficulties as a result of excessive expectations of accuracy and the failure of the majority to appreciate the true meaning of terms such as recurrence intervals, return periods or probabilities.

Forecasting

Forecasting is similarly beset with problems of heightened expectations because of advances in science and technology. Progress has been variable, as a result of differing ability to monitor conditions. Where monitoring is achievable with present technology, such as for the atmosphere or ocean surface, reasonable levels of forecasting accuracy have been achieved over recent decades. Satellite-based forecasts of the progress of tropical revolving storms (hurricanes, typhoons and cyclones) and the Tsunami Warning Service for the Pacific Basin represent the two pinnacles of achievement, and the rather less glamorous storm-surge forecasts also achieve good results (e.g. the Storm Tide Warning Service operated by the UK Meteorological Office). Even the generally available short-term (24 hour to 48 hour) weather forecasts achieve a reasonable level of accuracy (70–85%), although they are still incapable of providing information on the location, timing and intensity of small-scale phenomena such as individual tornadoes, thunderstorms, locally intense rain or hail, violent gusts, and so on. However, this information can increasingly be obtained with radar to yield short-term forecasts and adverse weather warnings. Because of these problems of scale, the normal approach adopted is to issue phased forecasts/warnings, which are up- or down-graded as increasingly precise information becomes available. In the case of geological/geomorphological phenomena (earthquakes, volcanic eruptions, landsliding, etc.) forecasting is much less precise because of monitoring problems, and hydrological problems (droughts, floods, avalanching) represent an intermediate position. Some of these points are considered further in Chapter 7 (pp. 127).

Risk communication

Risk communication continues to need improvement, as tendencies remain, at least in the developed world, to view environmental hazard impacts as inconvenient intrusions into "normal life"; as isolated and individual occurrences, rather than as part of a continuum in time, space and magnitude, and as problems for science and technology to resolve. Some of this may reflect the view, dominant in the 1950s to 1970s, that technologically advanced societies should be able to "control" adverse environmental conditions. However, appreciation of the limitations of science and technology, together with growing financial realism, has resulted in the realization that human societies must continue to live with hazard and, therefore, must seek to reduce risk to tolerable levels. This can be achieved only through better information, explanation and education.

Conclusion

Recent decades have witnessed considerable increases in both awareness of environmental hazards and scientific understanding of their generation and behaviour. This has provided ability to anticipate (i.e. to assume or realize in advance) the likelihood of occurrence, magnitude of impact and probable consequences, to anticipate (i.e. to prevent by prior action) consequences through the development of planning controls, engineered structures, warning systems, emergency action and financial planning, and to anticipate (i.e. to look forward to) future hazard events as part of planned recovery programmes. Although it is true that many major hazard impacts tend to reveal hitherto unknown dimensions of threat or vulnerability (e.g. the 1985 Mexico City earthquake (see Degg 1992b) and the 1994 Los Angeles quake) and new hazardous threats are continuously being identified (e.g. stratospheric ozone depletion, radon, landslide generated tsunami in the Pacific Basin, meteor impacts), natural scientists continue to be confident that anticipatory measures can considerably reduce losses and limit "surprise", especially in the case of small and medium-size events. The IDNDR represents a coherent attempt to raise the profile of anticipatory measures in hazard loss reduction, a philosophy that is well described in the phrase "Hazards are inevitable: disasters are not". Thus, in the example of natural perils management, the case for increasing the emphasis placed on anticipatory measures appears compelling.

HAZARD ENGINEERING

D. I. Blockley

Introduction

Two world-views of hazard, risk and safety with reference to the engineering industry are discussed in this section: the one technical and the other human and organizational. It will be argued that there is a need to unify these two views through the adoption of five key points, together with the definition of the new discipline of hazard engineering based on Turner's model of accident incubation (Turner 1978). The two basic activities of hazard engineering are hazard auditing and hazard management. The simple message that emerges is that safety depends upon good management.

Technical and human factors

It can be argued that there are two broad world-views of hazard, risk and safety, which may be referred to as technical/engineering and social science/management (Royal Society 1992).

The technical approach is often termed the "hard" approach (Blockley 1992a). It is typified by formal models and by the need for "objective" measurement. Its goals are clear, but there is often an underlying basic confusion between science and engineering. Science is concerned with accuracy and objective truth in the search for knowledge. Engineering, by contrast, is concerned with the production of artefacts, or systems of artefacts, in order to satisfy particular human needs. As noted in Ch.1, the technical theories of risk assessment and reliability have evolved from this paradigm and consist of scientific and engineering models of the physical aspects of a problem expressed usually in mathematical terms. The practices of risk, safety and hazard management vary greatly across the different types of the engineering industry. Thus, for example, the ideas of safety auditing are well advanced in the process industries, and reliability theory is well advanced in structural engineering.

There are limits to the technical approach that are often unrecognized. At the scientific level the developments in modern physics, in quantum mechanics and in the new theories of deterministic "chaos", have shown that there are distinct limits to what we can know. Likewise, in mathematics, the theorems of Godel have demonstrated the limit of formal languages.

31

The social science approach is broadly split into the individual (psychological) approach and the group (sociological) approach. It is often termed the "soft" approach in scientific discussions. It is typified by informal models, with few formal models and statistical measurements. Its goals are understanding through discussion. Theories such as the "Turner model" (Turner 1978) and those concerning human error (Reason 1990) are descriptive, but they nevertheless aid understanding.

Academic work (research and teaching) in engineering is dominated by the scientific approach. However, in engineering practice a tension exists between the demands of a scientific way of thinking and the practical need to deal with human and organizational problems (management). It is no coincidence that the training need most requested by graduate engineers is for management courses. Blockley (1992a) has described this tension in terms of the "Two Cultures of Engineering". The manifestation of the two cultures is clear in structural design, where, for example, is on the one hand there is the technical approach of reliability theory, with the vast amount of research that supports its development, and on the other hand, the managerial approach to quality, and hence safety, of quality assurance, about which there is very little research.

The 1992 report of the Royal Society has illustrated the marked division between the "hard" and "soft" approaches, where quite separate and apparently unconnected chapters deal with technical and human factors.

It is argued here that these two "world-views" urgently need to be integrated to a common purpose, recognizing that risk and safety are issues lying at the interface between the technical and the social. It is proposed that this could be accomplished by the adoption of five points as follows:
- the explicit recognition that hazard is a sociotechnical problem
- the use of "systems" thinking
- the generation of a culture of reflective practice, rather than technical rationality
- the preference of responsibility to reliability
- the articulation of limits to models.

Systems is a modish word that people are using more and more often, so what does it mean? It really is a subject or framework that facilitates the ability to think and talk about another subject; it is a meta-discipline whose subject matter can come from virtually any other discipline. Thus, it is more an approach than a topic, a way of going about a problem. In simple terms, the systems approach is one that takes a broad view, which tries to take all aspects into account and which concentrates on the interactions between different parts of the problem. Some of the key concepts in this approach are world-view, holons, hierarchy and appropriateness. These ideas are amplified in Checkland (1981), Blockley (1992a) and Senge (1990).

The reasons for the lack of systems thinking in science and engineering are largely historical. Schon (1983) argues that, over the past century, the professions have been successfully incorporated into the universities, but that they have paid a very heavy intellectual price for joining. His thesis is that universities are slaves to a particular epistemology, which he calls "technical rationality". He defines this as instrumental problem-solving made rigorous by the application of the scientific method. It is a view deeply embedded in our institutional context and it is a brave person who challenges it. The two cultures of engineering, discussed earlier, are a direct result of this approach.

Schon then develops the idea of the "reflective practitioner" but does not provide a definition. Blockley (1992b) has pursued this idea, defining reflective practice as the thoughtful achievement of practical competence within a theoretical framework that encompasses both science and engineering. According to this view, everyone acts according to some need or purpose. Scientists aim for knowledge with certain qualities. Engineers aim for artefacts with certain qualities. Certainty seems to be a basic human need, probably stemming from the desire for self-security (i.e. safety). Science is concerned with achieving zero risk through certain true knowledge – an aspiration now known to be impossible. There is an important distinction between the collective search for scientific truth and for engineering safety. It is that the objective of the scientific method is to reduce the problem into a closed world model where all relevant factors are identified and relationships expressed. This is not normally an option for engineers; the world is open in the sense that all relevant factors may not be known and the relationships between the known factors may be inconsistent and incomplete. Thus, although the possible solutions of an engineering problem are numerous, so are the possible futures. This is illustrated by the fact that, when engineering disasters happen, they are often the result of unintended consequences of human action. Interestingly, this is exactly what Popper described as the business of the social sciences.

The concept of the reflective practitioner captures the balance between reflection and action that can be identified in scientists, engineers, academics or other professionals. There is a spectrum of interests across the range from philosopher to road sweeper; all are important, all are contributing.

In line with the idea of the reflective practitioner, Blockley (1985) has argued that the technical/engineering concept of reliability should be replaced by the social science/legal/management concept of responsibility. The taking of responsibility implies not that one has earned the right to be right or even nearly right, but that one has taken what precautions one can reasonably be expected to take against being wrong. The responsible engineer is not expected to be right every time but he/she is definitely expected never to make ill considered or lay mistakes.

Thus, in a very real sense the concept of responsibility is more fundamental than that of reliability. It is embodied in the law of tort, in which the standard of care is that of a reasonable practitioner. The problem then is how to decide what is reasonable. Peer group judgement is usually advocated. The difficulty is that such judgements are often decided finally in the law courts. As Furmston (1992) has pointed out, the legal process is not the process through which safety can be improved or indeed lessons learned, since the objective of the courts is to allocate blame. Douglas (pers. comm.) has also pointed out that the allocation of blame is not a healthy sign if we wish to move to a safer society (see Ch. 3).

It is rare for engineers to articulate the limitations of the technical approach, particularly to their clients, since it is feared, with some justification, that to do so will result in commercial disadvantage. The lack of effective communication of risk between engineers and non-technical clients and between engineers and the general public is a major problem. No matter what approach is adopted to problem-solving it is necessary to construct models, whether formal or informal. There are at least six attributes of models that are worth exploring; function, grounding, form, specification, applicability and completeness. These are compared for science and engineering in Table 2.1, and definitions of the terms are given by Blockley (1992b).

Table 2.1 Qualities of scientific and engineering knowledge.

Scientific knowledge	Quality	Engineering knowledge
Prediction	Function	Fitness for purpose
Explanation		
Simplicity	Form	Environmental
Elegance	Economy	Impact, beauty, cost
Truth, content, consistency	Grounding	Dependability
Precision, clarity		
Abstraction	Specification	Appropriateness
Domain width	Applicability	Relevance
Relevance		Practicality
Closed world	Completeness	Open world

The balloon model of hazard

A hazard has been defined by Fido & Wood (1989) as "A set of conditions in the operation of a product or system with the potential for initiating an accident sequence". Turner's (1978) model of accident generation is based on the observation that most systems failures are not caused by a single factor and that the conditions for failure do not develop instantaneously. Rather, multiple causal factors combine and accumulate, unnoticed or not fully understood, over a considerable period of time, a period that constitutes the "incubation period" of failure. This all occurs in a sociotechnical framework. Within the incubation period various types of conditions can be found in retrospect. First, events may be unnoticed or misunderstood because of wrong assumptions about their significance: those dealing with them may have an unduly rigid outlook, brushing aside complaints and warnings, or they may be misled or distracted by nearby events. Secondly, dangerous preconditions may go unnoticed because of the difficulties of handling information in complex situations: poor communications, ambiguous orders and the difficulty of detecting important signals in a mass of background noise may all be important here. Thirdly, there may be uncertainty about how to deal with formal violations of safety regulations that are thought to be outdated or discredited because of technical advance. Fourthly, when things do start to go wrong, the outcomes are typically made worse, because of the tendency to minimize the danger as it emerges, or to believe that the failure will not happen. The incubation period, in which interlinking sets of such events build-up, is brought to a conclusion either by the taking of preventive action to remove one or more of the dangerous preconditions that have been identified, or by a trigger event after which harmful energy is released. The previously hidden factors are then brought to light in a dramatic and destructive way. There is then an opportunity for a review and reassessment of the reasons for failure, and adjustments can be made in order to attempt to avoid a recurrence of similar events in the future.

The notion of a sociotechnical system stresses the close interdependence between people and their social arrangements and the technological hardware they make and use. People and technology interact with each other and, over a period, change each other in complex and often unforeseen ways.

To appreciate this, it is helpful to imagine the development of an accident (failure, disaster) as analogous to the inflation of a balloon. The process starts when air is first blown into the balloon, at which point the first preconditions for an accident are established. Consider the pressure of the air as analogous to the "proneness to failure" of a project. As the balloon grows in size, so does the "proneness to failure". Events accumulate to increase the predisposition towards failure. The size of the balloon can be reduced by letting air out and lowering the pressure, and this runs parallel to the effects of management decisions, which remove some of the predisposing events and thus reduce the "proneness to failure". If the pressure builds up until the balloon is very stretched, then only a small trigger event, such as a pin or lighted match, is needed to release the energy pent up in the system. The trigger is often confused with the cause of the accident. The trigger is not the most important factor; the overstretched balloon represents an accident waiting to happen. In accident prevention, it is thus important to recognize the preconditions, to recognize the pressure in the balloon. The symptoms that characterize the incubation of an accident need to be identified and checked.

Hazard engineering

Blockley (1992a) has proposed that there is a need to identify a new activity called hazard engineering. This should be concerned with the identification and the treatment of exceptional circumstances where hazards exist that need to be controlled by specialist skills. The identification of hazard would be through hazard audits, and the control would be by hazard management.

For many projects, where the problems are relatively straightforward, the hazards may be managed, without the use of specially trained hazard engineers, in a way similar to those used presently. For more complex projects where problems may be anticipated, then the appointment of hazard engineers, either in house or as separate consultants, would need to be considered as part of the contractual arrangements.

Currently, there are many differing types of hazard, safety and risk audit systems in use. However, there are few, if any, that are comprehensive enough to include an adequate consideration of the sociotechnical interactions. Dester (1992) has outlined four types of existing audit: management, technical, unsafe acts and safety. The chemical industry, for example, has a well established record in the use of safety audits. The impetus to improve safety performance in the UK chemical industry seems to have been a report by the British Chemical Industry Safety Council in 1969, which described an investigation of safety practices in nine US companies. As a result, a guide was produced in 1973 in which safety audits were advocated as a means of assessing the quality of safety effort, which is not necessarily indicated by quantitative measurements of losses. This point was also made by the HSE in 1976, when it was stated that accident frequencies or incident rates are not a reliable guide to safety performance and it was suggested that systematic inspection be adopted. The 1973 guide listed five main elements in an audit:
- identification of risks
- assessment of consequences
- selection of actions
- implementation of actions
- monitoring of changes.

In 1991 the Chemical Industries Association revised the safety audits guide to include occupational health and environmental protection. Thus, the scope of an audit was allowed to vary in scale from a supervisor's inspection to a corporate site review, to cover any combination of safety considerations and to include any activity or systems of work.

Management audits using the International Star Rating System (ISRS), produced by the International Loss Control Institute in 1988, evaluates 20 major categories and for each category there is a detailed questionnaire. Scores are awarded for each answer, producing a rating for each category. The accumulated rating is a measure of an organization's safety effort. Examples of the categories are leadership, management training, planned inspections, task analysis, accident investigations, emergency preparedness, personal communications and off-the-job safety. This system has been used most notably on the Channel Tunnel project and by London Underground. Gaunt has reported that a study of 173 organizations using ISRS has shown positive effects in improved management skills, reduced accident rates, improved investigations and communications, but with negative effects of increased paperwork workload and time and effort. Another example of a management audit is "Construction Chase" produced by the Building Advisory Service and Health and Safety Technology Management Ltd in 1990. It is a self-examination and therefore there must be doubts about its effectiveness in the construction industry, where it is questionable that the safety culture is strong enough not to have external auditors.

Examples of technical audits are HAZOP and HAZAN, which are well known systematic critical procedures for identifying hazards. It is an audit of the design intention applied at various stages in the development of a new process plant or in a major alteration to an existing one. A description is given by Kletz (Blockley 1992a).

Dester (1992) has compared the extent of these various existing types of audit and has argued the need for a more general framework of ideas based upon a sociotechnical systems analysis. There is insufficient space here to describe the details of this framework, but it involves an assessment of evidence for proneness to failure from the three broad sources of the project itself, the state of the industry and the state of society. Under each of these headers a vast hierarchy of concepts have been identified. The evidence of proneness to failure is accumulated by examining particular attributes of each concept. For example, poor overall project management may be such evidence.

Some of the current research in this area is developing techniques of artificial intelligence for hazard engineering. In particular, work is being undertaken to produce a system that can "learn" from case histories of failure in order to identify the "proneness to failure" (Stone et al. 1989). In this work, case histories are captured as narratives, which are then translated into event sequence diagrams. Patterns of events are detected by using connectivity and discrimination algorithms on a computer, and groups of events are offered to the builder of the knowledge-based system as possible candidates for higher-level concepts. Event sequence diagrams are then merged to form higher-level stories. At the top level all stories merge into one. The system could then be used to measure proneness to failure by comparing the results of an audit with the accumulated information in the knowledge base (KBS).

The European Construction Institute have published a guide to total project management of construction safety, health and environment, which suggests a practical framework for hazard engineering (1992). It is clear that safety, risk and hazard are part of quality management, and so in direct practical management terms the best way of avoiding the disasters of the future is to create a caring and learning organization with improved teamwork, communications and mutual and self-regard. Senge (1990) has identified five characteristics of a learning organization: systems thinking, personal mastery, mental models, team learning and shared vision. By personal mastery he means people who are continually clarifying and deepening their personal vision, focusing their energies, developing patience, and striving to see reality objectively. They are in control and not drifting. By mental models he means the deeply ingrained assumptions, generalizations, or even pictures or images, that influence how individuals understand the world and take action. This is culture as defined earlier. These ideas are directly in line with those of Turner (1978), who said that successful design gives high priority, high status and high rewards to integrating all aspects of the resulting product, identifying and bringing out into the open potential clashes between requirements for different forms of specialized performance, and in taking steps to resolve such clashes. In very simple terms, this is good management.

Conclusions

- There is a need to integrate the technical/engineering view of risk and safety with the human/social science/management view.
- The five points suggested to accomplish this integration are: the explicit recognition that hazard is a sociotechnical problem; the generation of a culture of reflective practice rather than technical rationality; the use of systems thinking; the preference of responsibility to reliability; and the articulation of the limits to models based on the attributes of function, form, grounding, specification, and applicability.
- The Turner model of the incubation of an accident has been interpreted as providing a means of understanding a hazard as a set of preconditions for failure.
- It has been argued that there is a need for a new discipline called hazard engineering, which would consist of the two activities, hazard

auditing and hazard management.
* In the final analysis, safety depends on good management practice.

RESILIENCE, FLEXIBILITY, AND DIVERSITY IN MANAGING THE RISKS OF TECHNOLOGIES

David Collingridge

Introduction

Many writers, especially those advocating more incrementalist views on decision-making, suggest that the wise decision-maker ought to keep options open, because the future is so uncertain; the only certainty about it is that all plans will be overtaken by events and surprises, sometimes pleasant but more often painful. The recommendation to keep options open is clearly intended as a general rule. However, in the real world it is obvious that decision-makers constantly defy this suggestion, undertaking wars, radical technologies and political revolutions, where options are lost in copious numbers. As a description of how decision-makers cope with uncertainty, keeping their options open is falsified by every edition of a newspaper. Prescription, by contrast, is generally seen as of greater importance; it is far more important to suggest ways in which decision-makers might improve their practice than it is to describe their often pathetic attempts at making choices. Normative incrementalism holds that the future is so uncertain that any decision-maker will be advised to favour choices that keep future options open. This is not the end of the matter, of course, for maximum openness would be absurd; for example, building fossil-powered generating plant near every major town, because it keeps options open. Other elements must be traded against open options, but the central point is that, whatever objectives decision-makers might have, whatever preferences, needs, desires, values, or utility functions, they are likely to enjoy better returns for those decisions where future options remain open rather than from choices that close them off.

Flexibility and resilience

Several authors have attempted to operationalize the idea of keeping options open, or maintaining what can be called resilience. Unfortunately, there is no consensus on terminology, with results that are often confusing for the unaware reader. Fiering (1982) applied what he termed "resilience" to water resources, and operational researchers developed the concept of robustness for use in planning decisions (Gupta & Rosenhead 1968, Ackoff 1970, Rosenhead et al. 1972, Best et al. 1986). Eppink (1978) and Krijnen (1979) developed the concept of strategic flexibility to reflect the structure of an organization that allows it to cope with uncertainty. Economists such as Carlsson (1989) have attempted to incorporate considerations of flexibility in the standard theory of the firm, and both Sawhill & Silverman (1983) and Aggarawal & Soenen (1989) consider flexibility as part of investment decisions. Supporters of the evolutionary theory of economic change have placed uncertainty at the very centre of economic transactions (Nelson & Winter 1982). Ecologists have developed the idea of the resilience of biological systems (Sachdeva 1984). Strategic management researchers have also shown increasing scepticism that management's choices should be justified, recognizing that uncertainty cannot be eliminated (Allaire & Firsitotu 1989, Friend & Hickling 1987, Milliken 1987, Quinn 1980). This varied literature displays a healthy appreciation of the problems of uncertainty in human choice. However, the measures that have been suggested are either extremely theoretical and of doubtful use in real world choices, or else they are far from a polished finality.

Flexible and inflexible technologies

However, Collingridge (1980, 1983, 1992) has provided a concept of flexibility that is deliberately simple and straightforward, and has the additional benefit from the point of view of this discussion in being particularly suited to decisions about technology. Learning about options can only be through trial and error. Options that can be learned about through trial and error, quickly and inexpensively, are therefore more likely to yield the decision-makers the values they seek. Some options possess features that tend to delay learning and to make it expensive, and are to be avoided by wise decision-makers. For example, technologies that take a long time to be completed have long lead times that will slow up learning and make more expensive whatever mistakes might be made. With increasing unit size, in the search for scale economies, there will be less room for experimental comparison of several versions of the technology, again slowing learning. With technology of high capital intensity, mistakes are likely to be expensive, because the main costs are already sunk long before any problem has been identified. Lower capital intensity brings more ways of adjusting the project in the light of discoveries of unexpected mistakes or opportunities.

Infrastructure can be thought of as other technologies that are dedicated to the technical project in question. For example, a large single-purpose dam and lake impounded behind it are of no use for land irrigation unless there is a network of canals distributing the water right down to the individual fields. Likewise, the channels are of no use unless coupled to and fed by the reservoir. One only works with the other and has virtually no value in its absence, making the development of the reservoir and distribution channels activities that must be planned together and completed in a timely fashion. To underline the tight connection, the dam might be the technology and the irrigation channels its infrastructure.

Collingridge has termed technologies with these physical properties "inflexible" and claims that, because they are hard to learn about by trial and error, they typically perform very badly for those who choose them. Thus, no matter what values such inflexible technologies may have, those that employ them tend to lose heavily. Moreover, the concept can be of use in the decision-making process, because a project's flexibility can be assessed very early in its lifetime. Decision-makers should be persuaded to consider in place of a very inflexible technology, one that has a shorter lead time, smaller unit size, lower capital intensity and less dependence upon infrastructure. For example, tube wells are a much more flexible form of irrigation technology than large single-purpose gravity dams.

What is the relationship between Collingridge's flexibility for technologies and keeping future options open, or resilience? Collingridge clearly hopes that flexible technology will always promote resilience, but this remains to be shown.

Consider the relationship between a technology's capital intensity and resilience, which inevitably demands a few simplifications. It is necessary to compare production units with the same unit size, lead time and infrastructure, differing only in capital intensity. It must also be assumed that production costs are equal, or else the cheaper production unit will open up more options as to where to spend the money that has been saved. Let unit A have a higher capital intensity than unit B. The capital costs for A are greater than for B, but A's operating costs are less. If the plant fails to operate, for whatever reason, the operators cannot use the capital costs, which are sunk forever, but they do have control of the operating costs. The options open to the decision-makers are therefore to spend any amount of money up to the operating costs for the foregone units of output. The greater the operating cost, the more options that are open, making B more resilient than A. Sunk costs destroy options and reduce resilience.

Unit size will be considered next. For large units there are fewer arrangements of total capacity. Consider two production plants with the same capital intensity, lead time and infrastructure, but where the unit size of the first is double the second. There are always more ways of arranging the smaller plant to meet demand. The capacity of the small units can be adjusted to 1, 2, 3, units in contrast with 2, 4, 6, . . . in the large plant. Increasing unit size therefore inhibits resilience.

In the case of lead time, again consider production units differing only in lead time. Information about the technology's performance is acquired earlier where lead times are shorter and can be used to improve performance. Consider two projects where one will be completed no earlier than t_1, the other before t_2. For the first project it is possible to commence operation at t_1 or to delay this for any reason to t_2, or any later time. Remember that lead time is the shortest possible construction time for the project. However, the second project cannot become operational until t_2 or some later time. Thus, there are more options open for the first project than for the second; a short lead time therefore promotes resilience.

Infrastructure obviously reduces the options that are open. Irrigation from large gravity dam projects can only work when there are both the dam and the canals to distribute the water; neither has any significant value alone. Plans for one must necessarily include plans for the other, severely restricting resilience. Consider two plants differing only in that one has infrastructure G, the other G and H. There are more options open for the first. Running the second plant demands running both plants G and H, whereas operating the first requires running only G. Options of not running H are open in one case but not in the other. Infrastructure, therefore, tends to reduce resilience.

Promoting resilience

It is possible, therefore, to confirm Collingridge's hope that flexibility promotes resilience. If incrementalism is regarded as the insistence that uncertainties be handled by keeping future options open, then inflexible technologies provide a very easily researched counter-case, for their establishment breaks all the rules of maintaining resilience. This is why several researchers have defended incrementalism through a study of large technologies (Collingridge 1983, 1992, Morone & Woodhouse 1986, 1989).

However, resilience does not necessarily imply flexibility, because there are other ways in which the options open to decision-makers can be increased. Even with technologies, resilience may be promoted without considering flexibility. Thus, the use of diverse technologies of the same order of flexibility will make the final system more manageable as the future unravels, because its controllers will have more options available. For example, contrast an electricity-generating industry having only coal-fired plant, with one with a mix of coal and oil plant of the same lead time, unit size, infrastructure demands and capital intensity. The two sets of plants have the same flexibility, but the system consisting of only coal-fired plant has fewer modes of operation than the mixed system. If coal becomes temporarily unavailable, or if coal prices increase well above expectations, then there will be options of shifting electricity generation from coal to oil plants. This shows that flexibility does not tell the whole story. Flexible technologies are one way of increasing resilience, but diversity is another very familiar route to more options.

If there are two routes towards more options, what happens when they operate against each other? An important element in the publicized case for British nuclear plant in the 1970s and 1980s, was that it provided diversity of fuel supplies for a generation system that had been too dependent on coal. The case for Sizewell B claimed that diverse fuel supply from uranium and coal would give the electricity generators greater ability to cope with unexpected shocks of fuel price increases or strikes, the latter of particular importance remembering the strength of the miners' union at the time (Central Electricity Generating Board 1982). The case was that resilience would be increased by the adoption of technology that was inflexible. In reply, the critics of nuclear power argued that nuclear power stations are a very clumsy way to enhance fuel diversity compared to more flexible technical routes, such as buying coal from the world market, dual-fired coal–oil stations, gas-fired generating plant and so on, all of which increase diversity in ways that are more flexible and so easier to implement (Mackerron 1983, Thomas 1988, Collingridge & James 1991).

The case supporting Sizewell B maintained that fuel diversity would reduce generating costs by allowing fuel consumption to be adjusted to changes in the relative price of coal and uranium, a benefit that undoubtedly exists for coal- and oil-burning electricity plants. The British generating industry turned away from oil to coal following the dramatic increase in oil prices in 1973. Coal and oil plants have about the same capital intensity, around 33 per cent. Nuclear plant is more capital intensive, around 75 per cent of nuclear generating costs being capital. Little money is saved from not operating nuclear plant, making utilities use them as much as possible on base load. There would, therefore, be no hope of switching to more nuclear generation if coal prices increased, or to less if the price fell, for whatever nuclear plants were available would always have to be used flat out. The shift could only happen if more nuclear capacity were to be built, but with a lead time of 10–15 years such forecasts would have to be made a long time ahead. Shifting between coal and nuclear plant cannot, therefore, be rapid and smooth; it will be very lumpy and long term, threatening expensive errors if forecasts go adrift. Therefore, an inflexible technology, such as nuclear power, destroys options and reduces resilience. More flexible technology, such as coal and oil stations, allow rapid and less expensive substitution between fuels as market prices change. If decision-makers in the utility wish to operate their system with greater diversity, there are flexible and inflexible ways of doing this and the way that is flexible preserves more resilience.

Conclusions

This analysis has considerable bearing on decisions regarding technology and the risk management associated with them. Such concepts as flexibility, diversity and resilience have been developed by a research programme contrasting the unfathomable depths of ignorance in all-important decisions with our painfully limited abilities to acquire and analyze information. Decision-makers can never relax in the assurance that they have identified the very best option; any choice may be shown to be mistaken by future events that surprise the decision-makers. However, much research and propaganda on risk assessment and management assumes the very opposite; that some choices can be known to be the best and, therefore, do not require any humility from the decision-makers' search for resilience as a counter to deep uncertainty. In reality, it is necessary to admit that all that can be hoped for is a more or less efficient trial-and-error learning from experience of technology, and in this context ideas of flexibility and resilience become central to decision-making. Deep ignorance about the consequences of technology means that it is necessary to learn about them by trial and error, meaning high resilience. Two ways towards resilience have been described, notably flexibility and diversity. If flexible technologies surrounded by diversity are chosen, then it should be possible to control and adjust our technologies through whatever shocks and embarrassments the unkind future may have in store.

Risk management requires flexible technologies arranged with diversity. In the very early days of a technology's development, it is possible to consider its lead time, unit size, capital intensity and need for infrastructure; and, if it threatens to be highly inflexible, then decision-makers should consider ways in which flexibility might be enhanced, through shortening the lead time, or reducing the scale, capital intensity, or need for infrastructure. This is a central aspect of any satisfactory account of risk management for technologies.

CHAPTER THREE

Liability and blame: pointing the finger or nobody's fault

ABSOLUTIONISM VERSUS BLAME

The second key area of debate in risk management turns on the extent to which risk management regimes should be more or less "blame-orientated".

Those who favour a high-blame approach argue that effective risk management depends on the design of incentive structures that place strict financial and legal liability for risk onto those who are in the best position to take action to minimize risk. This principle has a long history in law and economics, particularly in discussion of the famous "Hand formula" in American law and its analogues (Posner 1986: 147–51). The claim is that, if liability is not precisely targeted on specific and appropriate decision-makers, a poorly designed institutional incentive structure will allow avoidable failures to occur. Without close targeting of liability, there will be too little incentive for care to be taken by those key decision-makers in organizations who are capable of creating hazards, and (the argument goes) "risk externalization" will be encouraged. Policies should, therefore, aim to support expanded corporate legal liability, more precisely targeted insurance premium practices, and regulatory policies that have the effect of "criminalizing" particular management practices and of laying sanctions directly on key decision-makers within corporations, rather than trusting corporations as undifferentiated legal persons (cf. Fisse & Braithwaite 1988).

The "blame" argument manifests itself in several ways. Some large business corporations build into their corporate safety policies a strategy for the dismissal of individual employees found to be responsible for safety violations. And some contributors to the risk management debate argue that avoidable accidents and failures may result from insufficiently individualized insurance (as in the case where government acts as its own insurer or where it introduces "insurance asymmetry" (Shrader-Frechette 1991: 88) by limiting third-party liability, as in the case of the US Price–Anderson Act limiting third-party liability of nuclear power plant operators to US$640 million), so that financial penalties for negligence are either non-existent or not discriminatory enough (cf. Perrow 1984). Others think that social efficiency in limiting risk externalization requires the ability to target legal blame on the designers and managers of a *system*, rather than on operators of its component parts, because blaming the operator will not necessarily create the appropriate incentives on designers and managers in relation to inherent system safety and system tolerance to minor human errors.

In recent years, a spate of transport accidents affecting UK citizens has led some jurists and other contributors to the debate to argue for changes in the legal system. They argue that it should be made easier to bring corporate manslaughter charges against those responsible for designing and directing organizational systems that are judged inherently unsafe, instead of focusing blame narrowly on an individual – for example, a driver or a pilot – who is not responsible for the broader corporate policies that create the setting for major system failures (cf. Field & Jorg 1991, Wells 1988, 1991, 1992). The sinking of the car ferry *Herald of Free Enterprise* at Zeebrugge in 1987 led to a corporate manslaughter suit against P&O Ferries being brought to court in 1990, partly as a result of pressure-group activity. Although that particular case collapsed, it may well be that further corporate manslaughter suits will emerge in the future, and such developments may significantly affect the incentive structure of senior corporate executives in relation to risk management.

In contrast to the strict liability approach, proponents of the opposing "no-blame" view are sceptical of the argument that a move towards more general "criminalization" of management or system design activity (rather than penalizing specific errors by operatives) will make for more effective risk management. Indeed, they hold that it may be ineffective or even counterproductive. Such critics point to other areas of policy where criminalization has led to the adoption of artificial legal devices designed to limit liability, rather than to produce real changes in behaviour. They argue that criminalization promotes tendencies to "go by the book", rather than the preferable more flexible approach, which adopts the most appropriate behaviour in the circumstances (Bardach & Kagan 1982); simply results in the export of risky activities to jurisdictions without criminalization policies, rather than producing overall reductions in risk; or may remove all motivation to undertake particular activities (e.g. if voluntary organizations withdraw facilities for children's play activities because of unaffordable insurance costs).

More positively, those of the "no-blame" persuasion believe that effective risk management means, on balance, a move away from mechanisms for pinning down blame after accidents. The assumptions are that a less "blamist" approach will lead to the provision of more information about malfunctions or bad practices and that such information will be promptly fed back to those responsible for future decisions. Although both of these assumptions are problematic, some major corporations (such as Shell International, which aims for a "no-blame culture"; Shell International 1988) base their corporate safety policy on such a declaration of principles, and the same argument can be found at the level of public management. Studies in the USA of so-called "high-reliability organizations" (LaPorte 1982, Roberts 1989, Roberts & Gargano 1989, Weick 1989; Sagan 1993: 14–28) have suggested that some complex systems can function efficiently only if all incentives to hide information about errors are removed, so that near misses and minor malfunctions can be fully analyzed and discussed in order to head off major accidents and failures.

Thus, those who favour a no-blame approach claim that in situations where the institutional process relating to major accidents (disasters) is primarily focused on apportioning blame, facts will be concealed or seriously distorted by the adversarial process, with negative consequences for risk management. If management paralysis and emotional responses in the media take the place of calm stock-taking in such circumstances, crucial information that could be relevant to learning will not be pooled. The result, so the argument goes, is that hindsight reviews, tougher corporate penalties and after-the-fact blaming processes will fail to deliver the resultant improvements in risk management that the blamists seek to achieve.

For example, following the prosecution of a pilot whose plane came close to colliding with a hotel near Heathrow Airport (London), there were reports of a "drying-up" of information provided by pilots on civil air mishaps and near misses, thus confounding the purpose of the reporting systems, which is primarily to increase safety by preventing similar accidents occurring. This reluctance to provide information in the face of possible prosecutions is exactly what those who favour a no-blame approach would expect when such reports are linked with blame and punishment.

As with the anticipation–resilience issue (see Ch. 2), few contributors to the risk management debate would put all the emphasis either on strict liability or on a no-blame approach. The real debate turns on precisely where the emphasis is to be laid on information and incentives in risk management. Those who incline to a strict liability approach think that legal and other blaming processes are inevitable and, at the margin, they are prepared to sacrifice the free-flow of *post-hoc* information in the wake of disasters or mistakes, in order to achieve strong enough incentives on managers and other actors to limit avoidable risks through the legal and insurance regime. From the opposite camp, the proponents of a no-blame approach argue that, since failures and near misses are inevitable, the opposite trade-off should be made in order to achieve maximum learning from failures as they occur.

Various aspects of this debate are considered in greater detail in the next three sections of this chapter. First, Celia Wells discusses how recent major accidents in the UK have come to the formal attention of the law courts amid increasing calls for business corporations to be charged with responsibility for the disastrous events that may overtake individual operations within their complex organizations. Such pressures for corporate responsibility are a clear expression of "blamism" in practice, and reflect the emotional needs of victims to give vent to outbursts of sentiment, the desire to reinforce morality and the view that the focusing of shame on those perceived as ultimately responsible for risk will result in improved risk management practices and procedures. She clearly shows how the faltering steps towards the development of corporate manslaughter charges are indicative of the complex relationship between disastrous outcomes, blame and the criminal process.

In the second contribution, Tom Horlick-Jones reflects on the complex interrelationship between the deep-seated human desire to gain retribution on perceived wrongdoers by the process of blame and the aspiration for effective risk management. Like the "high-reliability" school of risk management, he argues that "blamism" is unlikely to produce outcomes that are either efficient or just, and he outlines arguments in favour of the adoption of "no-blame" regimes.

Similarly, in the final section, Neil Johnston accepts the invitation to argue for a no-blame approach to risk management, basing his discussion on the belief that neither guilt nor blame have any legitimate part to play in a truly effective system of risk management. Using examples from the transport sector, he argues in favour of sanction-free systems of risk management, which emphasize collaboration, effective monitoring and voluntary disclosure. The growing success of such systems within the aviation industry points to the difference between "blame" and "responsibility" (and indeed the different shades of the term responsibility) and the internalization of risk so that it permeates all aspects of practice.

CRIMINAL LAW, BLAME AND RISK: CORPORATE MANSLAUGHTER

Celia Wells

Introduction

The relationship between blame, risk perception and the criminal process is inevitably complex. Identifying some of the influences that make up the intricate pattern of criminal law and practice involves critical questions about the role that blame plays in the construction of the social institution of criminal justice, and how perceptions of risk help to sustain the blame process. Blame may be functional for individuals as well as cultural groups. In contemporary Britain the demands for criminal punishment to satisfy feelings of vengeance appear to be growing. The purpose of this section is to link these observations to current preoccupations with holding corporations liable in criminal law following safety breaches resulting in mass death. First, attention will be focused on some ideas about blame and the role of criminal laws in modern society, before describing the legal background to corporate manslaughter. In the following sections, the concern is to emphasize the significance of the social construction of death in the recent development of prosecutions of corporations for manslaughter.[1]

Blaming corporations

Some writers assert that we are witnessing an increased tendency towards blaming collective institutions for the misfortunes that befall us, a trend reflected in both civil and criminal law (Bush 1986, Douglas 1992, Rabin 1992). Perceptions of corporate organizations and their responsibilities for mass death have, it is argued, undergone a change with less blind faith in the ability or willingness of corporate organizations to take safety seriously. Business corporations are increasingly expected to provide

1. I am extremely grateful to Mike Edwards and Tess Newton for assistance, generously funded by Cardiff Law School, with the research on which this paper draws.
   The essay is based on a paper I gave in a seminar series at the LSE in 1993 and I am indebted to Tom Horlick-Jones, the organizer, who continues to be the source of many references and ideas for my work. Because the essay was drafted some time ago, parts of it have now crept into some other publications, particularly Wells (1995a,b).

compensation for injuries that in earlier times would have been attributed to individual fault or fate (Bush 1986). This decline in confidence in major institutions, business and government itself has led towards more legalisation (Lipset & Schneider 1987, Giddens 1990, Galanter 1992, Horlick-Jones 1995), and in particular it appears that instances of corporate negligence resulting in death are more likely to be translated into calls for manslaughter prosecutions.

Debate as to whether corporate bodies could or should be liable for deaths caused through lack of regard to safety, needs to be conducted against a background understanding of the operation of criminal laws as they apply to individuals. It would not be sufficient for the debate to rest solely on that understanding, for there may be good arguments either that those laws are inappropriate for individuals or that they would be incongruous for corporate bodies. However, the dialogue will be more fruitful if it is not isolated from the insights that the social practice of criminal law can bring.

Because criminal laws and the criminal process are a familiar part of the institutions of state, it is easy to make assumptions about them. Commonly, law is conceived in instrumental terms, as a means to "protect" citizens from the behaviour of the lawless minority. A moment's thought reveals that this does not accord with the actual practice of criminal law and punishment. Not only are there many other means by which social control is achieved, but any attempt to draw a causal connection between criminal enforcement and reduced crime is fraught with problems. However, another view of criminal law is that it has an ideological function, that it makes statements about the boundaries of tolerated behaviour. In the case of corporate risk-taking, it is difficult fully to separate these two conceptions. And, although arguments about deterrence have fallen from favour as regards the punishment of individual offenders, different considerations may apply to corporate bodies.

The move towards blaming corporations for major disasters bears witness to the theory that blame generally, and criminal blame specifically, is used by people as a way of making sense of the world (Lee 1981; Taylor 1983: 107). There are two connected trends in recent writing about crime that help to underline the theme of blame that runs through this chapter. First, there has been a revival of interest in Durkheim's theory of the relationship between legal sanctions, social structure and public sentiment (Calavita et al. 1991; Garland 1991: ch. 2). The second is the renewed concern with ideas of vengeance and shame and their role in modern society (Braithwaite 1991). In Durkheimian analysis, punishment is a social institution that reinforces matters of morality. Punishment is neither rational nor instrumental; it is irrational and emotional. But it is also ultimately functional, in that giving vent to outbursts of common sentiment strengthens the social bond.

This ties with Braithwaite's argument that a re-integrative theory of shame is crucial to crime control. What is interesting here is the essential functionalism ascribed to shame and to vengeance. Durkheim saw law as either repressive or restitutive. Repressive laws inflict suffering and punishment, and penal laws are therefore characteristic of them. Restitutive laws seek to return things to as they were, and their sanctions are characteristic of civil laws. In their study of two dam disasters, Buffalo Creek in the USA and the Stava dam in Italy, Calavita et al. show that, despite calls for condemnation, in both cases the legal response was restitutive rather than repressive. Differences in legal culture meant that criminal proceedings were automatic in Italy, yet their repressive effect was reduced by the imposition of lenient sentences. Despite victims initially labelling the Buffalo disaster as "murder" and "criminal negligence", a grand jury eventually decided that no-one should be held criminally liable. Restitutive sanctions can emerge, the authors argue, despite an emotional reaction of outrage or shock (Calavita et al. 1991: 419).

Braithwaite (1991), on the other hand, uses the notion of shaming as specifically non-repressive. Shame can be used in a re-integrative rather than stigmatizing way. Forgiveness, apology and repentance need to be elevated to cultural importance, implying that restitutive sanctions can have a place in criminal law. It is a mistake, argues Braithwaite, to see shame as connotative of pre-industrial, folk society with clear networks of relationships. Modern communications systems may mean more interdependencies rather than fewer.

Using these ideas, this section explores the emergence, since the mid-1980s, of corporate manslaughter as a cultural and legal form. As corporations are brought within conventional criminal law enforcement patterns, the effect will be complex and fragmented. Any role that criminal law has in relation to safety will reflect and reproduce, as well as create, attitudes to risk (see Garland 1991: 252).

Legally constructing death

On the one hand, corporations are subject to the same criminal laws as any individual; on the other, their status as a juridical rather than a natural person poses difficulties in determining their responsibility. Those difficulties have been overcome in English law through a theory that those at the apex of a company – directors and other officers – act as the company when carrying out their duties.[2] Undoubtedly, more sophisticated theories of corporate accountability could be employed, but the

---

2. *Tesco Supermarkets Ltd* vs *Nattrass* (1972) Appeal Cases 153.

argument here is that the slow development of corporate manslaughter is attributable to wider causes than the narrow legal conception of corporate criminal liability.[3] Under discussion here is the application of the law of manslaughter to corporations, and to do this requires some appreciation of the breadth and elasticity of homicide law.

Almost all jurisdictions include unlawful homicides, the most serious class of offence. The structure, scope and sentencing implications of such laws vary across time and place. In comparison with many schemes, the common law division between murder and manslaughter as applied in England and Wales, is relatively simple, if fluid. Several conditions must be satisfied before a homicide will be regarded as unlawful. It must be caused by another person, which will include corporations as "juristic persons" but exclude "acts of God" or "natural causes".[4] Assuming cause, and in the absence of lawful excuse such as self-defence or prevention of crime, a death will amount to murder if accompanied by intention to cause death or grievous bodily harm.[5] Even where such intention is proved, the offence is reduced to manslaughter if it is provoked or the defendant's responsibility is diminished.[6] Manslaughter generally marks the border between, on the one hand, homicides to which criminal blame attaches and, on the other, those deaths regarded as accidental or to which no criminal blame attaches.[7]

Manslaughter is a hybrid category broadly covering recklessly caused deaths. A recent appellate decision specifies the circumstances in which deaths caused from breach of a duty of care may amount to manslaughter. The provision of a public transport service or other commercial enterprises would raise such a duty. If the breach of duty amounts to "gross negligence", a manslaughter prosecution could be instituted. In the Court of Appeal, it was suggested that gross negligence includes the following:

indifference to an obvious risk of injury to health; actual foresight of the risk coupled with the determination nevertheless to run it; appreciation of the risk coupled with an intention to avoid it but also coupled with such a high degree of negligence in the attempted avoidance as the jury consider justifies conviction; and inattention

---

3. An interesting development is increasing judicial recognition of a systems-based form of accountability (see Wells 1995c).
4. The problematic in such categorization is rarely acknowledged in legal discourse. For a history of corporate criminal liability, see Wells (1993a: ch. 6).
5. *R* vs *Nedrick* [1986] 3 All England Law Reports 1; intention can be inferred from defendant's foresight that death or grievous bodily harm was virtually certain.
6. Homicide Act 1957, §3 and §2 respectively.
7. Civil liability for negligence goes further and may be sought not only for those deaths that fall within criminal homicide but also for some deaths that fall outside criminal liability.

or failure to advert to a serious risk which goes "beyond inadvertence" in respect of an obvious and important matter which the defendant's duty demanded he should address.[8]

In a related appeal, the House of Lords affirmed that gross negligence was for the jury to determine by asking themselves whether, "having regard to the risk of death involved, the conduct of the defendant was so bad in all the circumstances as to amount to a criminal act or omission".[9] Of course, this leaves open wide scope for juries to establish the limits of acceptable risk-taking and was the definition of negligence used when a jury found OLL Ltd guilty of the manslaughter of four schoolchildren whose canoeing expedition, organized by the company, ended in tragedy in the English Channel in December 1994. Although it was the first corporate manslaughter conviction in England and Wales, the foundations were clearly laid in the unsuccessful prosecution of the ferry operator, P&O, for manslaughter following the *Herald of Free Enterprise* disaster in 1987, itself only the third such prosecution in English legal history.[10] Although the current definition is arguably more pliant than that prevailing at the time of the P&O trial, manslaughter has always been a broad category of homicide, which raises the question as to why so few cases have ever come before a jury to enable them to express their condemnation of corporate recklessness. This is obviously the kernel of the argument propounded in this paper: that an understanding of corporate manslaughter requires an open-textured analysis, not only of legal doctrine (including that of corporate liability as well as of manslaughter) but also of social and cultural perceptions of risk.

Socially constructing death

How do we account for the paucity of corporate manslaughter prosecutions, given that many of the 500 workplace deaths that occur each year are regarded as avoidable? Of 739 deaths in the construction industry between 1981 and 1985, the HSE estimates that over 50 per cent could have been avoided by positive management action (HSE 1988a). Not only were hardly any of these pursued as manslaughter cases, but also in only

---

8. *R vs Prentice and Others* [1993] 3 Weekly Law Reports 927, per Lord Taylor C.J., at 937.

9. *R vs Adomako* [1994] 3 All England Law Reports 79, per Lord Mackay, L. C. at 87.

10. There had been two previous prosecutions – the High Court held that a corporation could not commit a felony, as manslaughter then was, in *R vs Cory Bros* [1927] 1 KB 810 and the defendant company was acquitted in *R vs Northern Strip Mining, The Times* 2, 4 and 5 February 1965.

35 per cent of workplace deaths are prosecutions brought by the HSE (Bergman 1994) for any of the low-stigma health and safety offences for whose enforcement it is responsible.[11]

It would be wrong to attribute this pattern simply to a resistance to corporate or business liability for crime, for road deaths present a similar situation. Until recently, more than 5000 people died in UK road traffic incidents each year, yet few resulted in manslaughter prosecutions.[12] Although these deaths are investigated by the police, and not by a separate regulatory agency such as the HSE, there is a choice between charges under road traffic legislation or common law manslaughter. The statutory offence of causing death by dangerous driving now has a maximum penalty of 10 years' imprisonment, whereas for manslaughter the maximum is life. Until 1991, the ingredients for the statutory offence were exactly the same as for reckless manslaughter,[13] yet most were pursued under the former less stigmatic road traffic legislation. In 1990, over 300 people were sentenced for causing death by reckless driving.[14] The fact that these were not categorized in the official statistics, alongside the 607 homicides that year, indicates something about attitudes to different types of reckless conduct. In sentencing terms, the reckless drivers were also treated more leniently. The proportion of those given immediate custodial sentences is high in both the reckless driving and the manslaughter categories (71 and 80 respectively).[15] However, a significant number of the reckless drivers were fined (29) or given a community service order (33), sentences that were not used at all following manslaughter convictions.[16] The recent increase in the length and severity of sentences for serious road traffic offences is evidence of changing attitudes to road safety, but there is no doubt that a conviction for manslaughter still conveys a different message than one for causing death by dangerous driving.

It is helpful to place the issue of corporate liability for disasters in the context of these other deaths, in order to emphasize the significance of the social, rather than the legal construction of events. In particular, it is little

---

11. The average fine against companies in years 1988–90 was £1940 (Bergman 1991) and against all employers £1134 in 1991/2 (Health and Safety Commission *Annual Report 1991/2*; London: HMSO, 1992). The maximum fine imposable by magistrates (who hear most cases) for breach of duties under §2 to §6 of Health and Safety at Work Act 1974 was increased from £2000 to £20000 in 1992. Fines in the Crown Court are unlimited.
12. In 1991 the number was 5590 (*Social trends*,Table 7.23; London: HMSO, 1992). See also Spencer (1985).
13. Road Traffic Act 1991, §1, substituted "dangerous" for "reckless". Causing death by reckless driving was held to be synonymous with manslaughter in *R* vs *Seymour* [1983] 2 Appeal Cases 493 and *Jennings* vs *US Government* [1983] Appeal Cases 624.
14. *Criminal statistics* (London: HMSO, 1991).
15. See the discussion in Ashworth (1992: 105).
16. Because the statistics deal generically with all manslaughters, a detailed comparison is not possible.

appreciated that it may not occur to anyone that the consequences of a major disaster might come within the *legal* definition of unlawful homicide, especially if there is no particular individual as to whose recklessness it can be attributed. There is no automatic police investigation or prosecution, as would happen in Italy, for example.[17] A disparate range of factors, of which the legal definition of the offence plays a necessary but certainly not a sufficient part, determines whether a prosecution is brought, whether a conviction results (especially in the Anglo-American system, given the reliance on juries), and the severity of sentence imposed.

In order to come under the spotlight of a murder or manslaughter charge, killings have first to come to the attention of the police (or, in some cases, the Director of Public Prosecutions). Workplace deaths are rarely reported to the police by health and safety officials. The kinds of deaths that inevitably attract police attention are those that take place in bar, street or domestic brawls, or during robberies, burglaries, or in other stereotypical, individualized scenarios. Most of these are accommodated within a category of manslaughter, not relevant for present purposes, based on an unlawful and dangerous act. Deaths caused recklessly by individuals carrying out otherwise lawful activities are far less likely to result in manslaughter charges. Those that do are difficult to characterize, but some recent examples of convictions in the UK include:

- a man whose two daughters died in a fire caused when he dropped a cigarette during a drinking session
- an anaesthetist whose patient died during a routine eye operation[18]
- an anaesthetist following patient's death in routine exploratory operation[19]
- the train driver in the 1989 Purley rail crash who passed a red signal light[20]
- an electrician whose negligent installation of a heating boiler electrocuted a man.[21]

Not all those prosecuted are convicted; for example, a gas fitter was acquitted of manslaughter, for death from fumes caused by his faulty service to a boiler.[22] And two psychiatric nurses were cleared of man-

---

17. Italy is an interesting example because, like many Continental jurisdictions, it does not recognize corporate criminal liability, yet a major disaster will inevitably be followed by the prosecution, for negligent homicide, of senior company officials.

18. *Guardian*, 27 January 1990. Appeal against conviction was allowed in this case and that mentioned in *n.* 21 below on grounds of a misdirection in law. The combined appeals gave rise to a reworking of the definitional requirements for this type of manslaughter, see *nn.* 8, 9.

19. *The Times*, 31 July 1990.

20. *The Times*, 4 September 1990.

21. *Guardian*, 31 January 1990. This was his second trial, the jury at the first having failed to agree.

slaughter charges brought after the death of a patient in a scalding bath.[23]

Although unusual, it is not unprecedented for directors of a company to be charged as individuals in relation to deaths caused to employees or customers. Indeed, there does appear to be an increase in the number of prosecutions of both companies and individual company officers.[24] Two directors of a plastic company were charged with manslaughter when they negligently caused the death of one of their workers;[25] and three directors of a coach firm were charged with manslaughter of a teacher and 12-year old girl when a coach owned by their company overturned.[26]

These examples give a sense of some of the deaths that hover on the boundaries between the criminal and the accidental. There may be many other recklessly caused deaths that could fit the legal definition of manslaughter but which are never considered, because of a lack of fit with the social or cultural paradigm. The desirability or likelihood of a prosecution for corporate manslaughter, following transport or other disasters caused by management disregard of safety policies or precautions, are not matters that can be assessed from a purely legal standpoint.

Corporate manslaughter

It would have been inappropriate to discuss the evolution of corporate manslaughter without considering some of the wider issues that have been touched upon. There is no doubt that "corporate manslaughter" has become a culturally recognized phrase since the mid-1980s. At the time of the Aberfan disaster[27] in 1966 there was little if any talk of collective criminal liability. The trend towards responding to disasters in terms of corporate manslaughter seems to have begun with the capsize of the *Herald of Free Enterprise* at Zeebrugge in 1987. The reasons for this change are varied and complex, but they are not a result of any obvious differences in terms of legal culpability reference points, such as negli-

22. *The Times*, 23 July 1990.
23. *The Times*, 13 March 1981.
24. For example, Thomson Tour Operators were reported as being under police investigation following the death of a holidaymaker from carbon monoxide poisoning last year (*Guardian*, 3 July 1995).
25. One pleaded guilty and the prosecution accepted the not guilty plea of the other, October 1988.
26. *Independent*, 22 March 1990.
27. The Aberfan disaster is the worst landslide-induced disaster in British history. At around 9.15 am on 21 October 1966 a portion of Merthyr Vale Colliery tip no. 7 collapsed and 107 000 m³ of material moved down slope as a flowslide, penetrating the village of Aberfan and largely engulfing Pentglas primary school. A total of 144 people died, 116 of whom were children.

gence, neglect or recklessness (Wells 1995a). For example, the *Herald* "legal story" opens with the Sheen Inquiry's damning criticisms of the ferry operator, P&O (Sheen 1987). Yet there is little difference in the language employed in the Inquiry Report after Aberfan,[28] a report eloquent and unequivocal in its condemnation, not only of the National Coal Board (NCB) for failures that led to the disaster itself, but also for its attitude to the Inquiry:

> . . . our strong and unanimous view is that the Aberfan disaster could and should have been prevented.[29]

> However belatedly, it was conceded by the NCB that the Aberfan disaster stemmed from their failure to initiate any policy in relation to the siting, control, inspection and management of tips.[30]

Despite this clear censure of the NCB, there is no evidence of contemporary discussion of a possible corporate manslaughter charge. The Report itself does not connect its condemnation of the Board for negligence with the possibility of this having criminal consequences. Things have changed. It is clear from the P&O prosecution, from the abortive private prosecution for corporate manslaughter following the *Marchioness* Thames riverboat disaster in 1989, and the conviction of OLL Ltd, that corporate manslaughter now has a cultural as well as a legal meaning.[31]

However, there has been considerable institutional resistance to the translation of those meanings into an actual conviction, an example of law following, rather than leading cultural attitudes. The progression from disaster to trial was neither simple nor predictable after the *Herald* capsize. The Director of Public Prosecutions (DPP) only reopened the case after the inquest jury returned verdicts of unlawful death. The coroner discouraged a verdict based on corporate (as opposed to individual) manslaughter[32] and, when the trial eventually took place, two and a half years later, the trial judge directed acquittals on the dubious (but unappealable) ground that there was insufficient evidence of recklessness.[33]

As to the *Marchioness* tragedy, there was neither a public inquiry nor an inquest verdict to prompt the DPP into reconsidering his decision not to prosecute.[34] The report of the Department of Transport's Marine Accident

---

28. (Edmund-Davies LJ) Report of the tribunal appointed to inquire into the disaster at Aberfan, HC 553, HMSO (1967).

29. Ibid., para. 18.

30. Ibid., para. 178.

31. Evidenced by the Law Commission's discussion in its Consultation Paper 135, *Involuntary manslaughter* (London: HMSO, 1994).

32. His ruling that corporations could not commit manslaughter was appealed by relatives, *R vs HM Coroner for East Kent, ex parte Spooner* [1989] 88 Criminal Appeal Reports 10.

Investigation was not published until two years after the accident.[35] The justification for the delay was that it was necessary in order to prevent any prejudice to the trial of the Bowbelle's captain for failure to keep a proper lookout.[36] Two trial juries failed to agree a verdict, perhaps revealing an unwillingness to blame an individual, rather than the shipowners or the Department of Transport.

Meanwhile, in September 1990, the DPP had ruled out the possibility of charges against South Coast Shipping, owners of the *Bowbelle*.[37] A private prosecution was launched as soon as the DPP indicated that the case against the captain was being dropped. This prompted the DPP to take the almost unprecedented action of asking for papers from the private prosecutor's solicitor, with a view to taking over and dropping prosecution.[38] A week later, the DPP decided not to intervene after all, and the next day the MAIB Report was finally published. This rendered less credible the claim that it had been withheld previously for fear of prejudicing the captain's trial. An alternative interpretation, rather less favourable to the DPP, is that the Report was published at this point precisely *so as to* prejudice the private prosecution.[39] After further attempts by the defendant to prevent the prosecution,[40] committal proceedings finally began in June 1992, but the indictment was not made out.

Conclusion

More generally, this contribution has sought to introduce a perspective on the criminal process that begins to explore the relationship between blame and criminality. There is often talk of criminal proscriptions in

---

33. The trial judge's ruling that corporations can commit manslaughter is reported in *R vs P&O European Ferries* [1991] 93 Cr App Rep 73; the trial itself is styled *R vs Alcindor and others*, Central Criminal Court transcript, 19 October 1990. (see Wells 1993a). The re-introduction of gross negligence manslaughter, discussed above, was helpful in overcoming the problem raised by the P&O trial judge.
34. There are complex procedural reasons for the sequence of these public institutional responses, see Wells (1993a; 1993b: 47–51).
35. *Report into the collision between* Marchioness *and MV* Bowbelle *on 20 August 1989* (London: HMSO, 1991); it was completed in February 1990 and published on 15 August 1991.
36. 1988 Merchant Shipping Act, §32. No use was made of the offence under §31 of failing to ensure the safe operation of their ship against Bowbelle's owners. This was a new provision implementing recommendations of the Sheen inquiry into the *Herald* disaster.
37. An application for judicial review of this was rejected (*The Times*, 31 October 1990).
38. *The Times*, 3 August 1991.
39. The Marchioness Action Group certainly believed that publication was an attempt to block the private prosecution (*The Times*, 15 August 1991).
40. *R vs Bow Street Magistrate, ex parte South Coast Shipping Co Ltd* [1993] 1 All ER 219.

terms of territory or frontiers. But there is no determined terrain of "criminality", rather an "area of conduct to which, given prevailing interpretive conventions, formal proscriptions might plausibly be applied." (Lacey 1995). In a common law jurisdiction such as that in England and Wales, the development of a jurisprudence of corporate liability for crime, and specifically of corporate manslaughter, has been allowed to germinate, if not flourish, through incremental case law. The interesting questions are those that address the cultural, institutional and legal processes by which recent disasters have come to formal attention in the law courts. For risk management, an understanding of the contingency and unpredictability of those processes is as important as a comprehension of the law of manslaughter. As to the latter, the definitions are flexible enough to respond to changes in the cultural climate. If safety managers want to make themselves weatherproof, their barometers need to be tuned as much to the pressure of social constructions of accidents as to the legal categories into which they can potentially be placed.

THE PROBLEM OF BLAME

Tom Horlick-Jones[41]

Introduction

> Since the abolition of capital punishment, the British public has turned to those in charge during lurid disasters to sate its lust for retribution. Find someone to blame, cries the mob, and off runs Whitehall to offer up someone for lynching.

So journalist Simon Jenkins described processes of "ritual damnation" (*Sunday Times* 1989) following a string of disasters that took place in the UK during the 1980s, including those at Hillsborough, Bradford and King's Cross. Whether one considers the picture that Jenkins's rhetoric portrays to be an accurate one is a moot point. However, it does raise worrying concerns about the possible threat posed by the social impact of disasters to an efficient and just administrative response. But how real is this danger?

The wish to seek out and identify "who is to blame" is a common reaction to the powerful psychological shock of disasters, seeming to help people come to terms with their loss. This activity clearly has potential utility in the sense that the identification and punishment of wrongdoers may assist the prevention of recurrence. Indeed, some would argue (see the introductory section of this chapter) that precisely targeted liability "concentrates the mind" of key decision-makers, so providing an incentive system for avoiding failures.

In practice, targeting, as described above, is a problematic process. It is not clear whether it is a matter of identifying who is formally responsible for some event, or whether it involves extending a causal chain back to some defective, faulty or malicious action. Are there circumstances in which a simple accident becomes blameworthy? Can ignorance be culpable (Hacking 1986)?

In addition to the difficulty in seeking to target liability unambiguously, the process itself may have negative side-effects: discouraging learning processes within organizations and tending to promote strict rule-following behaviour.

In this section it will first be argued that "blamism", in the sense of precisely targeted individual liability, is unlikely to produce outcomes that

41. The author is grateful to Celia Wells for many useful discussions on the themes explored in this section. He absolves her of any responsibility (or blame) for his conclusions.

are either efficient or just. It will then go on to examine the advantages of introducing various sorts of "no-blame" regime, and the feasibility of such arrangements.

Targeting and the danger of systemic nets

Fundamental to the construction of criminal law in Western countries is a notion of individualism that sees actions as arising from the decisions of free, rational agents. Norrie (1991) and others have criticized this convention because of the implicit divorce of individual behaviour from social and power structures. Such difficulties are particularly stark in the case of disaster causation.

The targeting of sanctions in a "blamist" approach to risk management requires the identification of individuals who are, or were, in control of certain key decisions. Fitzgerald (1968: 122) has noted that there is a "general principle" which states that a person should not be punished for occurrences over which they could exercise no control. He observed that to penalize such an individual is not only unfair but also inefficient, "because it would not prevent similar occurrences in the future" (ibid.). If such targeting is to be implemented, then it must be accurate.

Disasters rarely arise from technical failures or "natural" hazards alone. They are sociotechnical events in which social, administrative and managerial factors tend to play major roles (Turner 1978, Horlick-Jones et al. 1993, Toft & Reynolds 1994). According to Turner's (1978) "incubation" model, minor failures accumulate, possibly over an extended period, perhaps years, to create an underlying causal chain. The effect of these factors, "a multiplicity of minor causes, misperceptions, misunderstandings and miscommunications" (Turner 1994: 216) weakens the overall system, but are individually insufficient to create a major failure. In this way, the preconditions are generated in which some apparently minor event can trigger a disaster.

In these circumstances the causal link between the trigger event and the "system failure" may be extremely complex. Wagenaar & Groenewold (1987: 596) describe such failures as "the consequences of highly complex coincidences". They recognize that, among the "multitude" of contributing factors, human error plays a "dominant role"; however, they conclude that the "stupid mistakes" that can lead to disaster only seem to be stupid with hindsight.

Human error, then, can trigger the disasters that ultimately occur for very complex reasons. In a sense, the impact of a simple error can be "amplified" by the sociotechnical context in which it takes place. In such cases of "complex causality" the context may not be clearly apparent to the individual who makes the mistake, leading them to being caught in what one might call a "systemic net" of circumstances beyond their control.

This problem has been recognized by Rasmussen (1990: 453), who has introduced the concept of "power of control", meaning the extent to which an individual determines the outcome of their actions. He recognized that:

Present technological development towards high hazard systems requires a very careful consideration by designers of the effects of "human errors" which are commonplace in normal, daily activities, but unacceptable in large-scale systems.

He went on to warn (ibid.):

There is considerable danger that systematic traps can be arranged for people in the dynamic course of events. The present concept of "power of control" should be reconsidered from a cognitive point of view, as should the ambiguity of stop-rules in causal analysis to avoid unfair attribution to the people involved in the dynamic chain of events.

The systemic nature of disaster causation calls into question commonplace notions of agency that play fundamental roles in the cultures of Western nations. These problems lie behind recent debates about the extent to which English criminal law is equipped to deal adequately with these events, and in particular with those conditions under which a corporation may be found criminally responsible (Wells 1993a, 1995b, c).

Organizations and failures

In disaster causation the action of individuals almost always takes place within organizational settings. This dimension complicates matters considerably, with the micropolitics of blame within the organization distorting both diagnosis of responsibility and processes of learning from past events (Jackall 1988, Sagan 1993, Toft & Reynolds 1994). In addition, following a disaster, the social and administrative environment in which an organization operates impacts upon the organization in complex ways, influencing its post-disaster behaviour (Douglas 1986, Bowman & Kunreuther 1988).

The relative roles played by individual workers, corporate structures and company directors in disaster causation have become an increasingly controversial issue in recent years (Wells 1995a). In the case of the capsize of the car ferry, *Herald of Free Enterprise*, outside the harbour at Zeebrugge in 1987, the official inquiry was clear in its recognition of the role of corporate factors (Sheen Report 1987: 14):

> The underlying or cardinal faults lay higher up in the company. The Board of Directorships did not appreciate their responsibility for the safe management of their ships (and) did not have any proper comprehension of what their duties are . . . From top to bottom the body corporate was infected with the disease of sloppiness.

Sir Jeffrey Stirling, Chairman of P&O, the parent company, responded to these charges by expressing the view (cited in Spooner 1992: 104) that:

> Although there have been discussions and talk and accusations about sloppiness in the management (of the company) . . . to suggest that they had a direct effect in that ferry capsizing in my view would be totally wrong . . . it gets a bit far-fetched that someone sitting on the shore should be hauled up for something not happening.

Despite Stirling's protestations, the evidence suggests that operating procedures that imposed critical constraints upon the actions of workers were inadequate. In mechanistic terms, the "power of control" of those at the operational end of the corporate body was structurally constrained by those higher up in the management hierarchy. In this way, power can be exercised structurally, rather than in the form of direct instructions or orders (Lukes 1974).

There is also evidence that in organizational settings individual workers do not behave according to the classical models of rational choice. It is not, as Hamilton & Sanders (1992) have recognized, a matter of agency arising from individual choice. Rather, they argue, authority frames the decisions in such a way as to produce a situation of role requirements and obligations.

Arguably, the responsibility of directors turns on the extent to which those at the top could reasonably have foreseen the possible implications of their decisions. In practice, of course, the links between strategic management and operational management are diffuse. Nevertheless, a cynical observer might comment that senior managers are often happy to accept the rewards of corporate success, while distancing themselves from failure. But if senior management is responsible for success, who is responsible for failure?

As Jackall (1988) has noted, transferring blame plays a very important part in the dynamics of corporate micropolitics, where it can be used to justify claims, cover up inadequacies, legitimize and bolster authority, and a host of other roles. Individuals who are blamed for corporate failures may have been seen as expendable or vulnerable to being "set up". Indeed, Jackall has discovered that one of the conditions for advancement within a corporation is the ability to avoid blame by the development of networks and alliances. In turn, management may seek to insulate itself from liability for failures by the production of unwieldy formal operating procedures, which, as Hale (1990) has found, may be unworkable in practical situations, so potentially compromising safety.

In his study on the limits of reliability for organizational management of high-risk military technologies, Sagan (1993: 278) recognizes the politics of blame as a key obstacle to organizational learning from past events:

> The safety regulations may have been poorly written, but it is easier for plant management to blame the operator, than to accept responsibility itself for writing incomprehensible rules or having poor review procedures. The cockpit switches may have been poorly designed, but it is cheaper to fire the pilot than it is to redesign the control panel. The captain's task may have required absolute perfection, but the ship's owners want the cargo delivered immediately.

When an accident occurs, he notes, investigators "round up the usual suspects: the control room operator, the pilot or the captain who committed an error". However, such behaviour is "extremely misleading", and blaming operators and thus protecting the interests of designers and managers will "increase the likelihood of future mistakes" (ibid.: 246).

Further evidence for the dysfunctional aspects of a climate of blame have been provided by recent events in the British civil airline industry. A confidential reporting programme ("CHIRP") exists in which pilots report their experiences of near misses and other potential disasters, so providing a useful learning process. Press reports (e.g. *Independent* 1991) suggest that, following the prosecution of a pilot whose aircraft flew dangerously near to an airport hotel, a significant "drying up" of reporting took place, with possibly serious future consequences.

The recognition that near misses and other failures are opportunities for learning about the behavioural characteristics of sociotechnical systems, leading to the possible avoidance of disasters, has led to proposals to establish "no-blame cultures" in organizations. Such approaches would seek to generate a climate of openness in which workers are not frightened to report minor incidents or unsafe acts, and senior management are receptive to critical ideas from lower tiers within the organization, customers and outsiders (Turner 1991, 1994).

An outstanding example of the implementation of these ideas is provided by the multinational oil company Shell. Since 1980 the company has adopted a series of management programmes designed to generate a corporate "safety culture" within which "no-blame" practices have a crucial role. The accident record of Shell's tanker fleet, expressed in terms of frequency of injuries, has fallen dramatically since the introduction of these measures in the late 1970s (*Lloyds List* 1994, *Seatrade Review* 1994).

So, "no-blame" approaches to risk management, contrary to being recipes for irresponsibility, seem to offer creative ways forward for managing complex sociotechnical systems within organizational contexts. However, the constraints placed on the establishment of "no-blame" culture by the micropolitical factors examined above, and by the structural features of the technology in question, pose a series of difficult management problems (see discussions in Turner 1991, 1994).

The social and institutional environment

Processes of blaming or not blaming within organizations may be significantly influenced by the organization's interactions with its operating environment. As Douglas (1986: 85) puts it, disasters "become enmeshed in the micropolitics of nstitutions". The feasibility of no-blame approaches to risk management within organizations may, therefore, turn on the sociocultural attitudes and the legal and regulatory practices of society as a whole.

The need to blame seems to be fundamental in a wide range of cultures and societies. The elaborate rituals documented in the classic anthropological studies of the Azande (Evans-Pritchard 1937) exemplify the search to allocate blame for a death or other misfortune, in this case by the utilization of certain magic oracles. Blaming, then, seeks to make some sense of the world, and to defend the tribe from future harm.

In the modern world, social psychologists argue in similar terms that blaming plays an important role in seeking to interpret, and to come to terms with, adverse events. However, they recognize that subjective factors can seriously skew this process of interpretation. The resulting "fundamental attribution error" (Fiske & Taylor 1984) is a tendency to blame undesirable events on individuals who are selected by their personal characteristics, without taking into account situational factors beyond these agents' control.

More generally, an influential group of cultural anthropologists, advocates of the so-called "cultural theory" of Mary Douglas and her collaborators (Douglas & Wildavsky 1982, Thompson et al. 1990, Douglas 1992), argue that blaming behaviour reflects the bias corresponding to distinct cultural formations or "ways of life". Blame is attributed, according to this theory, in such ways as to reinforce, and not challenge, existing attitudes and ideals.

In practice, therefore, for someone, or for an organization, to be considered blameworthy involves a process far more complex than merely demonstrating their causal role in some event. Although responsibility, in the sense of causal chains, is a necessary condition of blameworthiness, it is certainly not a sufficient one, the judgement involving a multiplicity of cultural, political, historical and other factors.

According to the cultural theory model, "individualists", a form characterizing entrepreneurial behaviour, tend to blame failures on bad luck or personal incompetence. In contrast, "egalitarians" tend to blame the economic "system", "society" or powerful institutions such as governments or big business. Indeed, Polisar & Wildavsky (1989: 152) go so far as to claim that:

> System blame, for instance, may serve egalitarians in their desire to discredit the unconscionable inequalities of markets and hierarchies.

They go on to argue that changes in American tort law may be explained by cultural shifts that have resulted in a tendency to blame systems rather than individuals. This theme is taken up by Wells (1995: 177–8) in her analysis of debates concerning use of the English criminal law against corporations. She recognizes "an undoubted shift, a change in the collective unconscious in relation to blame" and, significantly, that a corresponding: ". . . increased tendency towards greater legalisation has accompanied a decline in confidence in major institutions, business and government".

This conclusion relates directly to recent work by Horlick-Jones (1995) on the nature of disasters in the technologically advanced societies. Building on the work of Beck (1992) and Giddens (1990, 1991) on the role of risk in the "late modern" world, he argues that the concept of disaster has been socially constructed from traditional notions related to catastrophe, with their occurrence corresponding to the release of repressed existential anxiety. Modern disasters, reinforced by their media portrayal (Wilkins & Patterson 1990) as "explosions of outrage" (Horlick-Jones 1995), cause a perceived betrayal of trust by "expert" individuals and organizations (Horlick-Jones 1995, Horlick-Jones & De Marchi 1995).

The corollary of this argument is that blaming certain individuals or organizations may serve fundamental psychological needs, by re-establishing critical trusts necessary for people to cope with contemporary social life. The increasing complexity and reflexivity of technologically advanced societies may, therefore, lead to an enhanced tendency to blame.

In practice, such social, economic and technological currents weave a complex tapestry that forms and interacts with organizations in uneven and shifting ways. It has been recognized that the administrative means in modern societies by which responsibility is diagnosed and punishment dispensed reflects the influence of political, socio-economic and cultural factors (Garland 1990). Hewitt (1983) has observed that in the past the apparent meaninglessness of natural hazard catastrophes has been dealt with by "firmly locating blame". He goes on to recognize that "it would be quite naïve to imagine that the legacy of such ways of thinking does not still exert an enormous pressure upon our dominant institutions", of which the legal system is clearly a prime example.

The politics of blame: the Purley rail crash

A case example may be useful to illustrate some of the issues discussed in this section. On 4th March 1989, just outside the railway station at Purley, south of London, two trains, both heading for Victoria station in London, collided. Carriages from both trains were derailed and some rolled down an embankment. Five passengers were killed and 88 people, including three railway staff, required hospital treatment for their injuries (Department of Transport 1990).

The official inquiry by the UK Government's Department of Transport, then the relevant regulatory body, clearly identified the immediate cause of the accident as the failure of the one of the drivers to control the speed of his train in accordance with the signals, resulted in his train running into the back of the preceding train (ibid.).

The driver in question, who had a previously exemplary driving record, was prosecuted for manslaughter and sentenced to an 18-month jail term, with 12 months suspended, subsequently reduced to four months on appeal. The decision caused an outcry, with the rail union ASLEF describing the driver's actions as "an honest mistake", and threatening industrial action in response to "one law for the worker and another for the corporate body" (*Independent* 1990a, *Evening Standard* 1990).

Behavioural research had shown that the repetitive tasks involved in monitoring warning signals along the railway line could result in drivers getting into a "mind set" in which they believed they had performed a task, when in fact they had not done so. This research, which by the time of the trial had been recognized by the train operator, British Rail, was rejected by the judge as he did not think it was a contributory factor in the case (*Independent* 1990b,c).

The judge went on to state that (quoted in the *Independent* 1990c):

I have to look at the public concern that those who provide services to the public should do so carefully and (be) conscious of the implications of serious shortcomings such as yours in the performance of that service.

He went on:

Passengers put themselves in a very special sense in the hands of the driver. They trust him entirely. It is not just one person but hundreds who entrust themselves.

This position was echoed in the subsequent official annual report on railway safety (HSE 1991: 16):

Society places an onerous responsibility upon the drivers of trains to maintain constant vigilance. This was emphasized by the sentence of six months' imprisonment imposed at the Central Criminal Courts upon the driver of a passenger train who failed to observe a signal at danger and collided with another train.

In this case there was no suggestion that the driver had acted recklessly in terms of his own volition, or that his judgement had been affected by drink or drugs. The "mind set" research suggests that the driver, who could recall nothing of his actions, was not in a position of "power of control" for the omissions that led to the crash. In cultural theory terms, the "one law for the bosses" statement by the driver's union was a typical "egalitarian" interpretation of events; however, it is difficult not to feel some sympathy with the view that he was made a scapegoat for a much more complex failure.

There were, however, strong symbolic dimensions to the prosecution. In this regard, the judge's statement, quoted above, includes some important elements. Of particular interest is his assertion that the legal machinery needed to reflect "public concern", and his comments about the special trust relationship between driver and passengers. In addition, there is a suggestion that special standards of efficiency and care are expected in the work of someone who is the subject of so much trust.

In this way, as Garland (1990) has noted, the courts provide important ritual manipulation of symbolic forms, helping to structure contemporary discourses and thinking about blaming and deviance. Such penal rituals, he argues, function as a means of educating and reassuring their public audiences. In turn, legal institutions and their associated cultures are subject to the influences of "public opinion", which may manifest itself in a variety of ways, including internal political dynamics, overt lobbying and media coverage.

Arguably, the timing and circumstances of the Purley crash were particularly influential in this case. The accident took place just three months after the railway disaster at Clapham Junction in South London, where 35 people had been killed in another train collision. A climate of fear and concern about safety issues existed in Britain after the experience of several major catastrophes, including those at King's Cross,

Zeebrugge and Lockerbie, all within a relatively short period of time (see discussion in Horlick-Jones 1995, Wells 1995a). Additionally, until a few months before the trial, there had been continuing controversy over whether charges of corporate manslaughter would be brought against British Rail for its role in the Clapham disaster (*Guardian* 1989).

The Purley disaster clearly illustrates pertinent themes for studying the role of blame in risk management: the "power of control" of workers, the role of the company in determining operating conditions, the interpretation of events by the courts, the limits of the law and the possible role of a range of contextual factors that impact upon the behaviour of individuals and organizations. In this case the operator was not seeking to implement a "blame-free culture"; nevertheless, these events do raise serious questions about the extent to which such regimes could continue to function in the aftermath of tragedy.

Conclusions

The key conclusions to this section are not new. Nearly 30 years ago, Drabek & Quarantelli (1967) warned that to concentrate on personalized fault for disasters distracts attention from structural problems, and that understanding such problems may indicate changes vital to the welfare of society.

The establishment of blame-free corporate subcultures offers a constructive means of managing safety, efficiently and with justice, in a world of increasingly complex risks. However, organizations operate in social and institutional contexts, and the ability to maintain blame-free corporate cultures may be severely constrained by cultural and political factors. Moreover, social, economic and technological changes may be generating a greater need for blaming in an increasingly uncertain world.

In economic terms as well, "blamist" strategies of targeted liability are lacking in utility. Douglas (1992: 17), drawing on work by Calabresi (1970), argues that such targeting of blame is much less important for maintaining public safety than is the generous treatment of victims. "Paradoxically" she observes, it is cheaper on the "collective purse" to be generous to victims than to pursue litigation.

Blaming senior managers and directors for failures may be unjust in terms of the mechanistic involvement, or otherwise, of individuals in causal chains, yet it does target those most able to influence operating procedures. As Jenkins (1990) has argued, management may come to be increasingly vulnerable to the risk of legal sanctions unless seen to be taking all reasonable means to identify weaknesses in the sociotechnical systems over which it has control. However, the threat of such sanctions will produce compensating behaviour and, once again, will obscure understanding and block-learning processes.

The question of corporate liability is addressed in detail by Wells in the previous section. Advocates of criminal law reform (e.g. Wells 1993a, 1995a, c) claim that changes corresponding to a more realistic recognition of corporate liability would prove more efficient and just. Whether it would serve symbolically to restore trusts broken by catastrophe is a matter for speculation.

Ultimately, we may have to recognize that, despite our better judgements, we cannot avoid blame, and all blame-free approaches to risk management will at some stage be compromised, so necessitating a pragmatic application. The cultures of technologically advanced "late modern" societies have a fundamental, almost primitive, need to blame while possessing an unprecedented capacity to generate complex failures and disaster. Bauman (1993: 218) sums it up when he recognizes that:

Since what we do affects other people, and what we do with the increased powers of technology has a still more powerful effect on people, and on more people than ever before, the ethical significance of our actions reaches now unprecedented heights. But the moral tools we possess to absorb and control it remain the same as they were at the "cottage industry" stage.

BLAME, PUNISHMENT AND RISK MANAGEMENT

A. Neil Johnston

Introduction

Every accident, no matter how minor, is a failure of organization.
Professor K. R. Andrews

My task in this section is to argue for a no-blame approach to risk management. Although few ever feel comfortable with any argument taken to its extreme, I am certainly much happier promoting a no-blame approach than one that aims to seek out the guilty and punish them "*pour encouragez les autres*". I do not believe that guilt or blame have any legitimate part to play in a system of risk management. Why? Because the *management* of risk suggests that we seek, by various means, to control our exposure to risk and the consequences of human error. That, in turn, implies the creation of a system through which we will actively attempt to manage risk. An effective system will, necessarily, have many layers and employ multiple techniques (International Civil Aviation Organisation 1993, Orlady 1993).

Failures of risk management systems are frequently precipitated by individual acts or omissions. In the "system safety" scheme of things (Lloyd & Tye 1982) any inability to absorb the consequences of individual failure is ultimately considered to be a symptomatic failure of the system. The position I wish to argue is that an accident or serious incident ultimately derives from a system that is inadequately specified or designed, or which has insufficient "defences in depth" (Reason 1990: ch. 7). Immediate failures on the part of an individual are thus irrelevant for all practical purposes, save for the identification of essential changes to the system. Feedback on the efficacy of system performance must be the primary focus, given that it is the principal means of controlling risk. In this context, Captain Daniel Mauriño, Secretary of the International Civil Aviation Organisation Flight Safety and Human Factors Study Group, recently observed that;

> It is time to look at the systemic and organizational deficiencies which – by fostering human error – threaten the whole aviation system. No matter how well equipment is designed; no matter how sensible regulations are; no matter how much can humans excel in their performance, they can never be better than the system which bounds them. (Human Factors Revisited 1993).

From this perspective, punishing or blaming individuals will rarely play any productive role (Johnston 1991). Indeed, it may even serve to sustain or increase the exposure of the overall system to future risk, by virtue of reducing feedback about systemic deficiencies.

I start below by critically examining the notion of blame and its social role. I conclude by using case studies to examine the benefits of a no-fault approach to risk management.

Blame and responsibility

There are activities in which the degree of professional skill which must be required is so high, and the potential consequences of the smallest departure of that high standard are so serious, that one failure to perform in accordance with those standards is enough to justify dismissal.   Lord Denning (1978: 451B)

The captain of a ship or aircraft has ultimate legal responsibility for the safe transport of passengers and cargo. In each case the captain is the commander and his legal status imposes various duties and responsibilities. Historically these have often been equated to absolute responsibility – something to which the harsh findings of various nineteenth century Courts of Enquiry into shipping accidents readily testify (Barnaby 1968).

It can easily be argued that any operational accident involving a ship or aircraft *must* be the fault of the captain, given the nature of his onerous responsibilities and duties (Denning 1978). A consequence of this view is the tendency for those investigating incidents or accidents to reason backwards from the circumstances of an accident in the light of those rules and regulations deemed, *ex poste*, to be applicable. Such reasoning will inevitably find a stage at which the accident causal sequence could have been broken. This invariably is a point at which an individual failed to act in accordance with a general rule or regulation. Apportioning blame and responsibility for the accident or incident is then reasonably straightforward. Consider the following extract from an accident report:

3.37 Probable cause: The probable cause of this accident was the decision of the captain to continue the flight at low level towards an area of poor surface and horizon definition when the crew was not certain of their position and the subsequent inability to detect the rising terrain which intercepted the aircraft's flight path. (Aircraft Accident Report 1980).

On closer examination this probable cause statement turns out to be little more than a pejorative rewording of the actual circumstances of the accident (Vette 1983). It actually explains nothing about why this apparent act of folly took place, what the circumstances were, or what the captain and crew believed they were doing.

However, it does apportion blame. The crew, by virtue of not being in the correct location, was *ipso facto* deemed to be "not certain of their position". And by virtue of continuing to fly towards a mountain they could not see, the captain is further deemed to have made a "decision" to "continue the flight . . .". These are serious errors for a professional flight crew. The casual reader would think that the captain and crew actually knew they were in the wrong place. The investigating authority clearly felt that they should have known; after all, that was their duty and responsibility!

However, a subsequent Royal Commission of Inquiry determined that the captain and crew were entirely blameless:

393. In my opinion therefore, the single dominant and effective cause of the disaster was the mistake by those airline officials who programmed the aircraft to fly directly at Mt Erebus and omitted to tell the aircrew (Mahon 1981).

Justice Mahon, author of the Royal Commission report, immediately went on to add:

That mistake is directly attributable, not so much to the persons who made it, but to the incompetent administrative airline procedures which made the mistake possible.

Justice Mahon's findings gave an insight into the underlying systemic deficiencies that have to be addressed if similar accidents are to be prevented. Indeed, Justice Mahon's findings contribute to risk reduction and accident prevention precisely because they avoided apportioning individual blame and teased out the underlying causal factors. By way of contrast, it will be readily appreciated that the findings of the earlier investigation amounted to little more than a sterile statement of "pilot error"; it thus fails to contribute to future risk reduction almost in direct proportion to the degree to which it assigns blame.

Blaming the victim

Blaming the victim . . . consists of applying exceptionalistic explanations to universalistic problems.   (William Ryan 1976)

When considering cases where individuals are blamed for contributing to accidents or serious incidents, I normally start by using a "*substitution test*". This merely involves mentally substituting another actor from the same operational background into the circumstances of the accident and asking the question "In the light of how events unfolded in real time, is it probable that this new individual would have behaved any differently?" (Johnston 1991). If the answer is no, I then tend to the belief that apportioning blame has no material role to play, other than to hide systemic deficiencies and to blame one of the victims.

For example, in the case of the Zeebrugge ferry disaster, it is clear that the behaviour of the captain and crew on the occasion of the accident differed in no material way from that which had occurred on many previous occasions (Sheen 1987). This accident was precipitated by an individual who simply fell asleep – the single exceptional event. As the subsequent investigation demonstrated, this was "a disaster waiting to happen" (Harle 1994). All the central actors turned out to be victims of the circumstances in which they found themselves, given that each layer of safety protection had previously been subverted or rendered ineffective. Critical issues in this accident were the company subculture and working custom and practice – also important factors in the King's Cross Underground fire (Fennell 1988) and the Clapham Junction rail accident (Hidden 1989). A global response to such endemic deficiencies is manifestly essential if accident prevention and risk management is the aim. To achieve these objectives it is vital to transcend the narrow apportionment of individual blame.

But, in the public perception, accident prevention and risk management are not always the key issues, and perhaps this is the real reason why blame often tends to play such an exaggerated role. To blame a person is to label that person as having been in some way less than they should have been. The primary functions of this labelling process are social and psychological. Indeed, there is evidence that people tend to select the most blameworthy act as the main causal factor in the event of an unfortunate outcome (Alicke 1992).

For instance, when depressed economic circumstances in seventeenth-century England led to a breakdown in the informal system of community welfare, one social consequence was an increase in the perceived prevalence of witches in the community (Thomas 1971). Those marginalized citizens most in need of community assistance were often labelled witches and thus became further victims of a changing economic and social milieu. The dominant forces at play here were social and psychological; few would now accept that these evil witches really existed.

Blame: guilt and vindication

All punishment in itself is evil.   Jeremy Bentham

The key to understanding the role of blame is to consider it in social and psychological terms. Committing a serious error – especially if it has significant public repercussions – compels a reaction on the part of either government or corporate management. It is notable that such reactions often vary according to the degree to which the error was publicized or gave rise to public disquiet.

In such circumstances, punitive action is taken to be a signal of management's *intention* to act, and is often seen as a sign that management is *willing* to move to prevent a recurrence of similar events. Managers, especially if inexperienced, tend to feel they "have to be seen to do something" and a sanction of some form is the action that most readily comes to many minds. The manager's – or perhaps even the public relations department's – assessment of how senior management, or the public, expects them to act will probably be of greater significance in such circumstances than anything to do with justice, risk management or accident prevention.

Depending upon the circumstance, there may be a belief that perceived wrongs, or injury to innocent parties, should be publicly vindicated. There may be a need to demonstrate that the public will be protected in the future and that they will be rendered safe from further occurrences of similar errors, acts or crimes. When death or significant injury are involved, relatives often seek a means by which their grief can be publicly atoned. Apportioning blame can serve these social purposes while providing an opportunity for the wrongdoer to expiate his or her alleged guilt. Looking at it in these terms, the optimal social response by an alleged wrongdoer is to publicly accept his or her guilt and punishment, thus completing the circle.

All of these social issues are at play when, from time to time, legal sentencing policy is publicly debated. Such debates are normally precipitated by a perceived failure of the courts to act with sufficient decisiveness and retribution following an heinous criminal offence. I do not seek to deny the legitimacy of such arguments, but merely ask that they be seen for what they are and dealt with in the correct social and legal context. And, of course, my point is that the correct context has little to do with risk management.

Blame: retribution and deterrence

The beatings will continue until morale improves.
The Management

Blame plays another role. When we have identified and blamed a wrong-doer we legitimize retribution. As I hope I have already clarified, retribution primarily serves social purposes – although it is frequently justified in the light of its alleged role in preventing future acts, or omissions, of a similar nature (Error & Punishment 1991).

The notion that retribution serves such a purpose has always struck me as somewhat naïve (see Lederer 1979). It certainly appears to reduce ultimately to a very crude notion of human psychology. For instance, the idea that every time a physician makes an error of judgement he or she should be punished seems ludicrous, at least to most people. Most of us accept that such errors are far from intentional and that they result from imperfections of one kind or another – of education and training, inadequate information, undiagnosed disease processes, and so forth (although we are perhaps less inclined to forgiveness if it involves ourselves).

Physicians have some added advantages, not least that their worst mistakes normally involve single patients, who are subsequently buried! This tongue in cheek remark is not without its serious side. Accidents involving relatively large numbers of people will attract much more public attention than a consistently incompetent physician who quietly and unintentionally disposes of many patients.

In this context it is not encouraging to learn that autopsy reports suggest that physicians are in error on up to 40 per cent of occasions in their pre-autopsy assessment of the cause of death; however, for the purposes of the argument here, the key concern must be that physicians traditionally show little interest in feedback based on such information (McGoogan 1984). An added concern is the tendency for medicine to operate as a self-policing and opaque "closed shop". These issues having been raised, it does not seem immediately likely that any intervention involving blame and punishment is likely to succeed. The real issue must be how to manage and control medical risk. To do so successfully would appear to require appropriate structures, such as internal and external audit procedures, along with suitable feedback and remediation systems. Most important of all would be cultural and attitudinal changes on the part of physicians, including an acceptance of independent audit systems and a willingness to look critically at their clinical practice.

A punishment and blame ethos would actually serve to undermine and corrupt such initiatives. Indeed, it could be argued that medicine – as perhaps the ultimate closed shop – remains impregnable to reform precisely because of physicians' fears about the consequences of the professional *post-mortem*, or the unsolicited appearance of the legal profession. When under attack it is much easier to "pull the wagons into a circle" than to invite public scrutiny. Risk *is* controlled here, but it is professional risk that is ultimately controlled, not medical risk. The chosen means tend to be inward looking and do not lend themselves to transparent reform of medical practice. I disagree with such an approach, but I can understand it, and why it is deemed necessary.

This example can be used as a metaphor for any one of various professional and industrial settings and it illustrates that barriers to change in our risk management systems are both structural and cultural. Effective risk management systems can only operate effectively in a subculture that endorses and promotes feedback and remediation. A no-fault/no-blame ethos is clearly an essential element. Any such system must be also structured appropriately and be perceived to operate with integrity and effectiveness.

Learning from our mistakes

To err is human.   Cicero

Most are familiar with Cicero's observation on error, though fewer are aware that Cicero also said "only a fool perseveres in error". Implicit in the first statement is human imperfection. Implicit in the second is a desire to learn from, and minimize, that imperfection; this latter objective is exactly what risk management is all about (Hood et al. 1992). In practice it may even mean adopting a philosophical approach tantamount to "accepting error to make less error" (Einhorn 1986).

Risk management often involves sociotechnical systems. Technical aspects of such systems are readily amenable to quality- and risk-control techniques. The human side of the system, notably where it involves a dynamic exchange between humans and machines, is less predictable (Wiener & Nagel 1988). Furthermore, human operators act within a social system, including operational subculture(s) embedded within an organizational culture (Pidgeon & O'Leary 1994).

Organizations that are able and willing to respond promptly to feedback and modify their relationship to both the internal and external operations environments have been described as "generative" (Westrum 1992); such organizations can be contrasted with closed or "pathological" organizations (ibid.), complete with their blame culture, the absence of feedback, and employees who are adept at hiding errors and playing the politics of denial.

I consider below how to address risk management within a blame-free and no-fault climate by using various examples taken from the aviation industry.

Sanction-free collaborative risk-reduction systems

Blame, fault, mistakes, errors . . . these are words with negative connotations. Each readily generates denial, whether psychological or organizational. Denial is the enemy of rational behaviour, of change, and of constructive action. But denial mechanisms are central to human psychological make-up. Establishing a no-blame ethos may help prevent information on risk factors from being driven "underground", but it is rarely enough to ensure adequate feedback and proactive risk management. That requires suitable structures, a long-term perspective – and determined action.

Consider the dilemma faced by the US Airline Pilots' Association (USALPA) when it concluded that pilot alcohol abuse represented a potential health and air safety hazard. Medical provisions for the licensing of pilots mandate that alcoholism be permanently disqualifying. Most of the airlines approached by USALPA took the view that they did not employ any such pilots – and if they found one he would be fired! The Federal Aviation Administration (FAA), which issues pilots' licences, looked to the legal position and announced its hands tied. And, as is well known, alcohol abusers themselves elevate denial to an art form. Collectively, this constituted a lot of denial. During the early decades of aviation the problem of alcohol abuse was simply ignored and denied by everyone, thus cutting off essential feedback on aberrant pilot performance or behaviour. And, for as long as it was denied or ignored, overall risk within the aviation system was elevated to some extent.

And yet USALPA, in partnership with the FAA and several progressive airlines, eventually initiated peer administered mechanisms for the early identification and treatment of alcohol abusing pilots (Hoover et al. 1982). Early identification of addiction improved the overall safety of the aviation system, along with the chances of a successful recovery from the addiction. This was good for the pilots, for the airlines and for aviation safety. Each party gave a little, each received a lot; USALPA fully accepted that those members who could not cope successfully with treatment would lose their jobs. On the other hand, employers had to guarantee the re-employment of successfully rehabilitated pilots, and the FAA had to agree to re-issue their licences.

All parties had to accept USALPA's central administrative role and its *bone fides* on total confidentiality. In this no-fault/no-blame programme cooperation, successful treatment and abstinence guarantees a sanction-free return to licensed status and employment. The key objective is to identify, remove, treat and return "at risk" individuals, not to blame, label or punish them. Peer intervention and post-treatment monitoring are key identification and feedback mechanisms in this programme; it seems highly unlikely that this could ever work in any alternative intervention system. No-fault confidential pilot peer group activity (which originated in Canada) has spread beyond North America, and aspects are further discussed in Johnston (1985) and (Johnston & Kelly 1988).

Sanction-free error-detecting and reporting systems

The modern aircraft cockpit is characterized by a high level of teamwork, in which "error trapping" and "error management" are key priorities (Wiener et al. 1993). Nevertheless, an error-tolerant cockpit must be considered the *final* defence against random error, rather than the primary defence against systematic or design-induced error. If systematic or design-induced errors occur, these must be identified and addressed by appropriate means. Similar considerations apply to systematic or controllable errors arising elsewhere in the aviation system. Among the methods used to identify system deficiencies in practice are various no-fault feedback systems, three of which are briefly described below.

Operational monitoring   Operational monitoring is a sanction-free programme that promotes flight safety by providing airline management with timely feedback on the actual quality of flight operations. For these reasons it is an important safety programme (Lautman & Gallimore 1987).

Operational monitoring uses de-identified recordings of key aircraft in-flight parameters to monitor the "operational health" of actual airline flight operations. Recorded data from each aircraft are periodically extracted from the aircraft flight recorder. Collectively, this is used to assess overall operational "health". The information is also used to identify any individual flight parameters that exceed predetermined tolerances. Various precautions are taken to ensure the confidentiality of these data and, although agreements with pilots' associations vary, most operate using similar rules and practices. The only person who can access the identity of a particular crew is an agreed "ombudsperson", who is normally a representative of the pilots' association. Accessing individual data occurs only in exceptional and predetermined circumstances. Information obtained from operational monitoring may not be used in disciplinary action.

Two types of feedback on crew performance are available from Operational Monitoring.[42] The first concerns particular flights on which specific aberrations or "exceedences" beyond set tolerances occur. Only the "ombudsperson" can match exceedence information to a particular crew, and it is up to him or her to determine if it is necessary to interview the

---

42. Additional information can also be tapped from an independent "feed" to monitor technical integrity, engine condition and auto-land functionality.

crew to ascertain what took place. The "ombudsperson" feeds relevant information back to the operational and training side of the airline.

The second type of crew performance feedback relates to overall system performance. Adverse operational trends can be identified by cumulative analysis of the anonymous flight data. Appropriate feedback can then be provided to both management and pilots. Thus, for example, an airline might note that there is a growing tendency to land too far along particular runways, or that flaps are being selected at too high a speed in certain circumstances, and so forth. Given prompt feedback and accurate statistical assessment of the incidence of the particular problem, airline management can decide what type of remedial action is most appropriate, whether that be policy changes, remedial training, the dissemination of information, or whatever.

US Aviation Safety Reporting System (ASRS) Aviation has different sanction-free confidential reporting programmes operating within different countries and airlines. The largest of the national systems is the Aviation Safety Reporting System (ASRS) in the USA (Reynard et al. 1986). Any person involved in aviation can submit a confidential report to ASRS – including cabin crew, air traffic controllers and even passengers. By early 1993 the ASRS database held in excess of 150000 de-identified confidential reports and was receiving 3000 additional reports each month. Any researcher, and even members of the public, can request a printout of ASRS reports. The goal of ASRS is to improve system safety. This is achieved through the identification and rectification of safety problems, using feedback, analysis and a proactive system of communications.

Incoming reports are analyzed by a panel of specialists, mainly comprising retired airline pilots and air traffic controllers. They record key "codes" on the reports to assist with subsequent information retrieval. Reports are "de-identified" before entering the permanent ASRS database. This ensures that there is no information that might allow the identity of participant(s) to be deduced. Immunity from Federal sanction is granted to pilots and others who report an infringement of US Federal Aviation Regulations. (A method of achieving this, while ensuring full database confidentiality, has been developed.)

The primary focus of the ASRS programme is human error. The major objective is to identify risk associated with human error across the entire aviation system. Periodic alerts are issued to the relevant persons and organizations regarding identified areas of elevated risk (Chappell 1994). Many analyses and research reports have been prepared using ASRS data. Because of anonymity and immunity, ASRS receives many more reports than any other safety reporting system and the quality of ASRS information generally helps policy-makers address the root causes of errors (the "why") rather than the symptoms (the "what").

Voluntary disclosure ASRS data is provided by individuals. Another US Federal programme, called "Voluntary Disclosure", can be used by organizations to disclose information about occasions on which employees infringe operational rules or regulations (Federal Aviation Administration 1992). The objective is to ensure that the Federal authorities are voluntarily made aware of the relevant circumstances and to ensure that acceptable action is initiated. Voluntary disclosure protects the organization and offending individuals from ensuing punitive action. Although this programme is sanction-free, it is normal for some prescriptive remedial action to be jointly agreed with the authorities for subsequent implementation by the organization. Such action would normally be aimed at the prevention of similar acts in the future.

In concluding, it is important to note that Voluntary Disclosure does not avoid apportioning *responsibility* for acts or omissions. This emphasizes a most important issue, namely that instituting blame and sanction-free reporting/remediation systems does not mean that responsibility for error should not, or cannot, be allocated. Indeed, as implied by the Voluntary Disclosure programme, it may well be the case that the most effective remedial action will centre on ensuring that managers and individual operators fully understand – and are willing, and able, to discharge – their responsibilities.

Conclusion

> In the effective organization, then, the person lower down in the hierarchy is encouraged by explicit management word and deed to think and take appropriate corrective action, even if that action means admitting a mistake. In contrast, ineffective organizations often prevent inquiry because it might conflict with vested interests. Admitting a mistake or pointing out problems often is political suicide, since retaliation is sure to follow. (Westrum 1992)

The key to optimal functioning in any risk management programme is feedback regarding the quality of system operation and active management of "safety health" (Reason 1991). This mandates suitable and credible structures. The information obtained must be used proactively to assess areas of real or potential risk. As will be clear from the foregoing, the priority is always to ensure that feedback is relevant, valid, timely and accurate, in order that prompt and enduring action may be initiated. In the safety systems described above, priority is given to receiving such information, even to the extent of eschewing the allocation of blame or sanction. Those involved in such risk management activities normally accept that the overall integrity of system function is best assured by open lines of communication, combined with proactive structures and processes (Maurino et al. 1995).

The designers of such risk-reduction systems are invariably convinced that action leading to individual blame or sanction will adversely interfere with the quality of feedback, ultimately leading to a decrease in total system safety. The success of these systems is normally associated with the willingness of line management to accept some reduction in their power and operating autonomy, in order that accurate feedback on system functioning can be obtained. The particular success of the aviation industry in promoting such programmes is closely related to the strong safety imperatives that permeate all aspects of aviation practice.

CHAPTER FOUR

Quantitative risk assessment and risk management: risk policy by numbers

THE EXTENT TO WHICH "STATISTICS ARE SIGNS FROM GOD"

A third major issue in contemporary debates over risk management turns on the extent to which management regimes should rest on quantified evaluations of risk (QRA) as opposed to more qualitative assessments. The majority of writers and practitioners continue to defend the role of QRA and thereby uphold the spirit (in more subdued form) of Prior Roger Schulz of Taize, who claimed that "statistics are signs from God". Their challengers, on the other hand, are likely to be of a more cynical persuasion and to argue that there are "lies, damned lies, and statistics".

The emphasis placed on quantitative techniques of risk assessment undoubtedly reflects human preoccupation with rendering the future calculable and knowable, at least to some degree, thereby reducing feelings of helplessness. Knights & Vurdubakis (1993: 730) comment that by "constituting something as a statistically describable risk makes possible the ordering of the future through the use of mathematical probability calculus". As a consequence, "by creating a possibility out of what had been a threat, it enables us neither to ignore it nor to be frightened by it" (Turner 1994: 146).

It is most certainly true that much of the running in risk management policy has been made by quantificationists. The argument for quantification is that any rational system of risk management must rest on systematic attempts to quantify risks and to assess them against a pre-set array of objectives by methods analogous to cost–benefit analysis (e.g. the 10^{-9} failures per hour standard for flying control systems in modern aircraft). QRA has developed into a major instrument of public policy (The Royal Society 1983). Rigorous quantification of risk, it is held, is the only effective way to expose anomalies and special pleading (cf. Breyer 1993) and in that sense promotes policy rationality, for example by pointing to the very different value-of-life settings implicit in different areas of UK transport policy, notably road and rail transport (cf. Jones-Lee 1990, Evans 1992).

The technical sophistication to which the QRA approach lends itself fits well with legal and bureaucratic requirements for standard operating procedures, and the approach has been systematically adopted by bureaucratic organizations (Mitchell 1990). The approach remains the backbone of "rational" risk management in the UK, particularly in areas of complex sociotechnical risk and for many types of natural hazard. Supporters of quantification argue that there is no real alternative to QRA as the primary tool of resource allocation in corporate and public management.

However, there are important shades of opinion among those who favour an emphasis on quantification in risk management. Few practitioners of risk analysis would put all the weight of policy resolution in risk management on ever-more refined approaches to QRA. Many who favour a quantificationist approach stress the importance of understanding the causes and characteristics of different types of risk and not simply of establishing probabilities (important though that is). It is also widely conceded that QRA has several limitations in practice and needs to be combined with other, broader forms of information and analysis (e.g. in qualitative techniques of risk identification that feed into or complement QRA), while still maintaining that QRA offers an essential tool for promoting rational risk management and exposing key policy questions (cf. HSE 1990b, Reason 1990, Brogan 1991). Rather more worrying for the quantificationists is the assertion that the mathematical basis of risk is disputed (see Turner 1994) and that current theories of risk and probability are by no means uncontroversial.

In the smaller and much less influential opposite camp are gathered together those who are uneasy about placing heavy emphasis on QRA in risk management (cf. Wilpert 1991). Those of this persuasion point out that the assumptions involved in some QRA procedures are often both value-laden and implicit. They are sceptical of claims to be able to quantify risks with very high degrees of accuracy, particularly where changing human behaviour can make a crucial difference (e.g. where responses to safety measures or human-induced environmental change defeat predictions based on extrapolations of past data; see Adams & Thompson 1991). Lave & Malès (1989) have argued that no single decision framework can cater for all the relevant values that come into play in risk regulation, and that cost–benefit analysis and risk–benefit analysis, although scoring high in terms of economic efficiency relative to other approaches to policy, typically score low on the values of equity, administrative simplicity, public acceptability and risk reduction. Such scepticism does not necessarily mean outright dismissal of all attempts at quantification. More commonly, it involves a different judgement as to what the ideal balance should be as between QRA and other sources of information and judgement. It may involve interest in ideas of modifying and extending orthodox QRA techniques, or giving the approach "extra vitamins" by the inclusion of explicitly qualitative elements that might eliminate, or at least compensate for, some of its more obvious deficiencies (e.g. in the temptation to avoid inclusion of values of infinity in the analysis because of their mathematical intractability).

Some radical critics take up an even more extreme position and argue that QRA is not merely limited but actually harmful as a tool for risk management. Their claim is that QRA, although convenient for organizations facing public attack for their handling of risks, tends to exaggerate the ability to quantify risks reliably and may direct attention away from "safety imagination" for rarely occurring, hard-to-quantify areas (Toft 1990). In effect, the argument is that QRA may make the risk management system more vulnerable to the Type III errors of Raiffa (1968), where faulty specification of problems leads to the formulation of real solutions to what turn out to be the wrong problems rather than wrong solutions to the real problems. Hence, QRA becomes a "fatal remedy" (Sieber 1981) through mechanisms such as placation and functional disruption. This view (which is closely paralleled in critiques of overreliance on economics in other areas, such as that by Gorz 1989) holds that QRA's calculative techniques are not simply neutral decision aids but actually define the way that problems are perceived and addressed.

The radical critics' position is a minority one, and few practitioners take this view. But risk management practice often involves interaction between quantitative and qualitative techniques. An example of the latter is HAZOP (Hazard and Operability Study), a procedure in which a team of engineers and managers carefully consider the possible consequences of a range of malfunctions of each component in a proposed system, as well as reviewing safety aspects of start-up, shutdown and maintenance requirements. HAZOP has been widely used in chemical plant design to facilitate the identification of risks associated with the operation of a system outside its intended limits (Kletz 1986, Chemical Industries Association 1987), and its derivative GENHAZ has been proposed for the assessment of risks associated with genetic engineering (RCEP 1989, 1991). Much of the debate is, therefore, about how the two approaches can feed into one another, rather than simple advocacy of one or the other.

These positions are examined in rather more detail in the next two sections. First, Adrian Cohen, a former member of the HSE, reasserts the value of QRA as an essential basis for risk management. This is followed by an essay by Brian Toft in which he raises objections to some of the fundamental assumptions on which QRA is based.

QUANTITATIVE RISK ASSESSMENT AND DECISIONS ABOUT RISK
An essential input into the decision process

A. V. Cohen[1]

Introduction

This section, which necessarily reflects my personal views, has two main aims. First, to show that, in the sense of an attempt numerically to estimate a risk to life and health arising from industrial activity, quantitative risk assessment (QRA) is, where available and appropriate, a necessary input into risk decision-making, and secondly to discuss how such decisions can be formed.

In the terminology of the 1983 Royal Society report on risk (Royal Society 1983), QRA is a measure of risk *estimation*. This is only part of a wider process of decision, which includes risk *evaluation* – the consideration of the significance or value of identified hazards and estimated risks to those concerned with, or affected by, the decision. The terminology is developing; and the UK Health and Safety Executive (HSE) published a discussion document (HSE 1995) on the terms and concepts involved. But whatever terminology is adopted, although QRA is a *necessary* input into a decision, it evidently cannot be a *sufficient* or a *determining* input.

QRA can serve many purposes. It has an important role in the design of plant and equipment, in prioritizing possible safety modifications to plant, and in assessing the coherence and balance of the safety approach adopted. It can assist employers in demonstrating a satisfactory level of safety, for which they are legally responsible. The Management of Health and Safety at Work Regulations 1992 (HSE 1992e), for example, require at regulation 3(1) that "every employer shall make a suitable and sufficient assessment of the risks . . . for the purpose of identifying the measures he needs to take to comply with . . . the relevant statutory provisions." For large-scale or complex plant, the associated Approved Code of Practice (ibid.) notes, at paragraphs 13 and 14, that this could require some kind of QRA. Although a more judgmental risk assessment is appropriate "for small undertakings", the assessment of some risks, particularly at intermediate levels, might require "the application of modern techniques of measurement".

---

1. The author is most grateful for the helpful comments made by Mr J. D. Rimington, then Director-General of the HSE, Dr J. Le Guen and several other former colleagues in the HSE, as well as by Professors C. Hood and D. Jones of the LSE. The opinions expressed in this contribution are of course the author's.

Even without a specific QRA, reference to an exposure limit or professional design code can implicitly involve some kind of generic QRA and an implicit value judgement on adequate degrees of safety. Particularly in the past this choice has often been made by experts. Blockley (1992a)[2] notes that "codes are the means by which acceptable risk criteria are set without explicitly stating what those risks are". This section will discuss the growing need for such procedures to be explicit, and publicly accountable.

Decisions about risk can be at many levels and stages, and involve experts and others. So can many other decisions. Thus, for (say) traffic or tax law, at one extreme, decisions are made out in the field: at the opposite "strategic" extreme, it is for ministers in Parliament. What is peculiar to risk is the essentially technical nature of risk quantification, with its associated uncertainties, the way in which some hazards reflect high technology for which legitimate differences of value *about the technology itself* arise, and the necessary interaction between the technical and the political. These call for decision forums that can comprehend the technical issues, without becoming so committed to technocratic values as to lose credibility with the rest of society. This is a demanding, but surely not an overwhelming, problem.

The nature of risk quantification and its uncertainties

Even if risk quantification were as "precise" as, say, a measured weight, it could not determine decision because:
- the decision is essentially political: whether a hazard arising from one person's activity should put another at risk
- the risk may impinge differently on different people
- people have differing standards, values, etc.

But risk quantification is nothing like as precise as a measurement of weight. The estimates contain uncertainties, and often imply expert value judgements, e.g. "conservative" cautious pessimism. The existence of value judgements, and the nature of the uncertainties, must be made explicit. The range of uncertainties can be seen in three types of industrial risk to the workforce or to the neighbouring population.

Statistics of *industrial accident* involve problems of definition: of population covered and of type of accident. Thus, the HSE statistics[3] distinguish injury to employees, to the self-employed, and to the non-employed arising from somebody else's work activity, thereby including in the totals

---

2. A series of articles by various authors; the quotation in the introduction comes from Blockley's preface.

for the latter many accidents in the playground and the old-age home, for which there are varying propensities to report. Such variation will be greater for non-fatal than fatal accidents, often for quite innocent reasons, because of further definitional problems in reporting requirements. Attempts to deduce trends in precisely defined categories of accident will often involve small numbers. Interpretation will be beset by problems of fluctuation, and sometimes by tendentious comment.

For *diseases, fatal or otherwise, caused by exposure to toxic and other chemicals in the work environment*, there are additional problems in identifying the disease as industrially caused, and the causative agent, and in choosing a suitable maximum exposure limit. These issues are significant when a disease appears many years after exposure or when small increases are suspected in the incidence of a common disease.

Prediction of likely incidence from various exposures (the "dose-effect curve"[4]) is then complex: animal experiments and human epidemiology have to be combined. Chapters 3 and 4 of Royal Society (1983) and of Royal Society (1992) and, in more general terms, paragraphs 24–5 of HSE (1995) consider problems in this kind of quantification. One is unlikely to get unambiguous indication at the low doses to be met in practice. Experts can advise if there is likely to be a safe dose: "a concentration averaged over a reference period . . . at which there is no evidence that [the substance] is likely to be injurious to employees if they are exposed by inhalation day after day to that concentration" – the "occupational exposure standard".[5]

When the experts believe there is no such safe dose (the view taken of many carcinogens and of radiation), the appropriate framework of factors and bounding parameters adopted ("control geometry") is that of the "maximum exposure limit" or "dose limit" respectively: a dose that, with qualification, may not be exceeded (Carter 1989), and must further be controlled at a level that is as low as reasonably practicable (ALARP) – a concept that has existed in safety law for many years.[6]

A dose-effect curve extrapolates observation and research, and expresses expert opinion. But maximum exposure limit is set not by

---

3. Relevant statistics up to 1985–6 are available in the annual HSE's *Health and safety statistics* (London: HMSO). Subsequent statistics may be found in the *Annual reports* of the Health and Safety Commission (London: HMSO).

4. Otherwise known as a "dose–response curve".

5. *The Control of Substances Hazardous to Health* (COSHH) *Regulations*, 1988. See also, for example: HSE EH40/93, *Occupational exposure limits* 1993 (London: HMSO); J. T. Carter, Indicative criteria for the new occupational exposure limits under COSHH (*Annals of Occupational Hygiene* 33(4), 651–2, 1989).

6. A judicial definition of "reasonably practicable" may be found in *Edwards vs National Coal Board* (1949) (1KB 704 at 712, (1949) 1 All England Law Reports 743 at 747, CA, per Asquith LJ).

technical experts but by decision-makers – in this case in connection with regulations, proposed by the UK Health and Safety Commission (HSC). They will judge, by reference to what is practicable to do and to measure, a level that can be regarded as the maximum tolerable for the time being.

The problems are greater for what is normally described as QRA: the prediction of the frequency and consequence of *"major events"* i.e. disasters of various sizes. The output of the QRA is twofold: first a contour map of levels of individual risk, and secondly an expression of societal risk, usually as an "FN curve" depicting the predicted frequency with which various numbers of casualties will be exceeded. Events such as those killing more than 100 or 1000 people are rare. Therefore, direct statistics cannot be employed. Instead, the experts base their predictions on the observed or estimated frequency of various possible (and normally contained) causes of failure of major plant; for example, a faulty valve.

This must raise the following questions:

(a) Has the expert team thought of every route to failure?

(b) For each route, how do the probabilities of causes of failure reflect reality? Some will rest heavily on expert judgement. Some critics, therefore, call these estimates "subjective" (Elms 1992, Pidgeon 1992), although "judgemental" would seem a better word.

(c) Are the various routes to failure independent or interdependent (have the "common mode failures" been identified)?

(d) To what extent does expert judgement include "conservative pessimism" and to that extent overestimate risk?

(e) How is "human-error" taken account of?

The last of these questions is crucial. The wide variety of possible "human errors" ranges from individual lapses, to what HSE calls, at paragraph 10 of its "Human Factors in Industrial Safety", the "organisational characteristics which influence safety-related behaviour at work" (HSE 1989a). Operator "slips" and "mistakes" occur moderately frequently. Their observed frequency can be (but is not always) used in a QRA to predict further routes to disaster. Observed average frequencies of causes of failure will often include events arising from typical operator slips, and so on, and to that extent will take account of them.

What is not so easily predicted, and will almost certainly be omitted from a QRA, is the incident caused by mistaken priorities, or, for instance, the wilful laying aside of safety procedures. HSE notes that for many incidents where individuals seem at fault, the "fundamental failures were rooted deep in the organisations where the incidents occurred" (HSE 1989a: para. 33).

In the Chernobyl disaster the "team in control of the reactor deliberately removed layer after layer of protection provided by the designers, in order to complete a test" thus leading to "serious instabilities due to inherent flaws in the design of the reactor" (HSE 1989a: 2).[7] They thus "violated the operational rules intended to prevent such a situation". Moreover, they seem to have departed from their own test programme. The Report of the Inquiry into the Flixborough disaster (Parker 1975) describes this as "caused by the introduction into a well designed and constructed plant of a modification which destroyed its integrity". One can see similar situations described in detail in HSE (1989a), and find similar causes in some rail crashes (Nock 1966).

Those conducting a QRA must assume that normal standards of organization, management and inspection prevail, and must remind the decision-maker of this. A good example of what should be said is to be found in the conclusions of HSE (1991: para. 238):

> We cannot emphasise too strongly the importance of good managerial practice; a lapse from it will rapidly result in significant deterioration, and in risks much higher than the estimates given here, which could thereby become no more than paper studies, no longer reflecting reality.

Elms (1992) notes that the predicted frequency of structure failure can be lower than reported experience, and counsels caution in comparison, because of possible omitted routes to failure or human error. In contrast, for chemical plant, HSE (1989b) found an agreement between prediction and experience within an order of magnitude. (Comparison cannot, of course, be made for predictions of very infrequent major events, which might involve more than 100–200 deaths). Some factors, such as those noted in (a) and (d) above, will work in opposite directions. HSE noted that the agreement was "perhaps fortuitous" but "strengthened confidence in the power of QRA to make a realistic estimate" (HSE 1989b: para. 33).

Other hazards, wider in nature than harm to the person, such as global warming, ecological risks, and so on, show, even more clearly, limitations to quantification and to human knowledge (see Jones, this volume).

To sum up, expert estimates of risk are subject to problems of definition, interpretation, uncertainty, exclusion of possible routes to failure, and implied "values", which must be made explicit at each level of decision. They are also subject to tendentious misinterpretation in either direction. Judgement on what weight to give to the estimate, whether the implications are acceptable, and whether risks need to be reduced, is the decision-maker's, and not the expert's.

---

7. Also, see information compiled by the USSR State Committee on the Utilization of Atomic Energy on "The accident at the Chernobyl nuclear power plant and its consequences", for the IAEA Experts' meeting, 25–29 August 1986, Vienna.

How is risk quantification used in decision-making?

The decision should take account of the risk estimate, but cannot be determined by it. The limits of the tolerable region are set by comparison (not necessarily equation) with other risks. Moreover, the decision-maker needs to make as explicit as possible the framework and basis for decisions, and if necessary to subject them to public debate. In this respect, the HSE's papers on the tolerability of risk from nuclear power stations ("TOR") were a great step forwards (HSE 1988b, 1992a). Rimington notes that the concept of *tolerability* "depends upon our ability to quantify and compare risks in such a way that *people and governments can make informed judgements about them*, and to decide in particular whether the available benefits are worth the risk" [my emphasis]. This "provides a basis for legitimizing social and political decisions to accept significant risks from which some people receive greater benefits than others" and "involves a statement of the standards and limits that society undertakes to apply as part of this process".

QRA cannot by itself set such standards technocratically. Reid (1992) rightly criticizes attempts to do this. Most of us would agree. It is the job of society itself and its institutions to make that judgement, as both versions of the HSE's tolerability paper remind us (1988b: para. 2; 1992a: para. 9), as do paragraphs 55 and 58 of HSE (1995).

The decision might be formed at several levels, and in different organizations, many having their own technical advisers. It might be internal to a firm, or involve an accountable public safety or planning authority, a planning or other inquiry, or, for instance, government itself for the largest risks – consider for example the successive levels of decision in the Canvey risk assessment (HSE 1978, 1981).

All this implies two different, although interacting, functions. This corresponds to current practice, once a risk is expressed, rather than buried within some internal code. The nature of this difference of function and, in particular, the objectivity of the technical expert's advice, is now sometimes questioned. This issue will be discussed later.

The "*control geometry*" of a decision involving risk regulation is likely to be that noted earlier, and described in HSE (1988b, 1992a; 1995: paras 49–55). A risk must be below a certain level, and be further reduced ALARP, unless it is so small as to be broadly acceptable (tolerable). Wider considerations, and factors additional to risk, will apply if the decision involves issues such as "do we want this project: or this industry?".

HSE (1992a: para. 169) noted that "a risk of death of around 1 in 10000 per annum is the most that is ordinarily accepted by substantial groups of workers" – and proposed this level for individual risks to radiation workers. Thus, the HSE proposal is based explicitly on contemporary UK norms. But as radiation-induced deaths occur many years after exposure, and the delay has not been time-discounted, the proposal effectively sets a tighter standard for radiation than for other industrial hazards. HSE also proposed 1 in 10000 per annum for deaths to exposed members of the public (ibid.: para. 173), and following the Hinkley Point Inquiry, set a benchmark level of 1 in 100000 per annum for those living near new nuclear power stations (a modification following open public discussion of a proposed level). For societal risk, proposals were made by observing the "maximum [non-nuclear] calculated major societal risk we are prepared to tolerate" and then insisting for several reasons, on a tighter standard for nuclear plant (ibid: paras 180–89).

HSE (1989b) examined 16 cases, only two nuclear, in which QRA had been an input to decision. Each individual risk was well below the 1 in 1000 figure – presumably the practical effect of the ALARP principle. For ten cases in which societal risk was a significant decision factor (ibid.: iv) it was "impossible to specify upper limits of tolerability applicable to all risks . . . there are inevitably too many factors involved in assessing a societal risk, many of them involve supporting qualitative judgements and different ones apply in different cases." The paper found 41 such factors involved, including economic benefit, public confidence in the authorities, and so on. The institutions of society were thus making their own judgements, as indeed they should be, and using QRA as only one input.

HSE drew conclusions from this (ibid.: iii, iv). They can be summarized as:

- QRA is an element that cannot be ignored in decision-making about risk, but the numerical estimate must be treated with great caution.
- There are other important components of safety assessment, including human behaviour at all levels in an organization; this does not mean that QRA cannot assist judgement.
- It is not legitimate to "read across" risk by assuming that a risk tolerated for one type of hazard will also be tolerated for a very different hazard "so as to infer some uniform level or limit to be applied . . . particularly in terms of societal risk".
- It is important to predict what might occur, and keep the change of this as low as possible. "For this QRA is an indispensable element, but one to be used with caution and not to be applied mechanistically to demonstrate compliance with legislative requirements" (ibid.: iv).

The decision process: a discussion

The Royal Society (1983: 175–6) noted the difference between technical risk estimation and political risk decision, although it saw interaction as necessary:

The scientific expert will need to ensure that his views are taken account of and not misunderstood in government and intergovernment discussions. Equally, he should not expect decisions to be taken on scientific grounds alone. The government decision will be essentially political, although informed scientifically. We do not believe that we can construct a scientific process that by-passes an essentially political decision.

This was seen then as the bridge between the two sides of the risk problem. The 1992 report, which is published "not as a Report of the Society, but as six independent chapters", is not so explicit. Chapters 5 and 6 of that report (Royal Society 1992), and some articles in (Blockley 1992a) raise issues that can be summarized as:

- Certain kinds of risk estimation, particularly of major events, are said to rest so much upon expert *judgement* that they cease to be truly objective, and might be regarded as comparable with estimates of the perceptions of those exposed.
- Risk estimation, despite the "different functions", is regarded as effectively applying a technocratic standard. Implied value judgements can thus "pull" the decision.
- Society, therefore, needs to devise more appropriate forums to discuss these issues.

Are expert estimates useful, even with their qualifications?

Are the risk estimates, with their associated uncertainties and omissions, so "fuzzy" as to be useless, or so based on experts' opinions, even after the strict probing of a suitable decision forum, that their objectivity is questionable (Blockley 1992a, Royal Society 1992)?

The answer must depend on one's view of reality. I believe there is a distinction between, on the one hand, psychological and social studies that inherently describe a subjectivity – the *views, feelings, and so on of people and groups* – and, on the other hand, the results of QRA, which aim, however imperfectly, to describe an *underlying objective physical reality*. Perhaps my own original discipline (physics) has formed my attitude.

This raises the question of *who* will be using the estimates. Technical experts produce risk estimates, in technical language, for a variety of purposes: first for their own "design" purposes stated at the beginning of this contribution. It must not be forgotten that the world is a much safer place because they do this. These estimates are essential factors in judging risk tolerability and acceptability. The *risk regulator* will need to be *able* to receive and act on them. The public will expect regulators to be expert in this wider sense. This calls for a significant numerate experience and out-look. Indeed, it is hard to see how a risk-based regulation or issue could be formulated, without some language, standards or canons of proof that the technical experts have in common. Decisions on such issues should not be based solely on technical advice, nor are the UK decisions discussed in HSE (1989b) so based in practice. Moreover, the public expect a regulator to say publicly *what* the risk standards are, *what* the conclusions are, and *how* they were reached. Decisions so reached are not technocratic; they are technically informed, and publicly accountable.

The ultimate locus of many risk-associated decisions is not the regulator but some other authority, local or national, as HSE (ibid.) have shown. Those making the decision are then more likely to be laymen, who will need to take account of experts' technical estimates, and the views of risk regulators.

Do technocratic standards apply in effect although not intention?

Previous practice – for instance, the expert-drafted codes referred to in Blockley (1992a) – may have amounted to technocratic setting of standards. But public concern about "major" and "dread" hazards has led to explicit quantification. Decisions as to tolerability and standards are then made by public representatives or officials, in publicly accountable forums. This is becoming usual for other risk issues too. The process cannot be called *directly* technocratic.

Nonetheless, the Royal Society (1992: 153) refers to "*ex cathedra* pronouncements from a small remote group of experts", and calls for a more participatory style of decision about risk. This is an accusation of remoteness, and (perhaps *indirect*) technocracy, that pays only lip-service to serious qualitative arguments against a hazard. One can presume two hypothetical grounds for supposing this:

- The expert proponents of a project, or the technical advisers to lay decision-makers, might produce estimates that are so opaque or subtle as to be able to call the agenda and dominate a decision that should be more generally based.
- As a kind of converse, the decision-makers might be so wedded to general concepts of quantification as to give too much weight to the technological view.

Against this, technical experts do attempt to inform the public, not to mislead them. Moreover, the findings of HSE (1989b), suggest strongly that *in practice* the current institutions of society, in the UK at least, are well able to cope with the situation. The criticisms, therefore, may reflect not what is happening in practice, but concern about *what could happen*. The antidotes to this are:

- appropriate presentation of qualitative arguments by their proponents, and their fair and careful consideration
- open publication of risk estimates in comprehensible terms: easier said than done, but increasingly attempted by HSE
- suitable decision forums.

What forums can most effectively discuss risk estimates?

The *central problem* for strategic-level risk decisions is how to use an expert but often complex and imperfect assessment in a complex decision process that cannot, indeed should not, be made exclusively by technical experts, and which needs to be transparent and generally participative.

Not to use QRA would be to drive blind: to be moved by organized pressure groups with their own agendas, or by people's *uninformed* fears. QRA, and the views of scientific experts generally, are thus essential inputs into a risk decision that must be "informed scientifically", as the 1983 Royal Society Report stated explicitly, and which the institutions of society accept in practice. But that 1983 view implies the existence of appropriate discussion forums, and an effective interaction between expert and decision-maker.

Should the QRA itself be produced from a broader perspective?

Funtowicz & Ravetz suggest in Chapter 7 that when issues are at stake and uncertainties high, or there is a "multiplicity of legitimate perspectives", "extended peer communities" with "complementary expertise whose . . . affiliations lie outside that of those involved in creating or officially regulating the problem", might "perform quality assurance and critical assessment" of proposed solutions, or draw attention to other possible consequences.

This view raises many questions. Would the "complementary" experts be self-appointed? If appointed, then by whom, and by what criteria? Would some decline to participate? How would discussion be structured? Is the aim limited to a "developing discussion on the technical aspects" that would "positively enrich the processes of scientific investigation", or would it also aim at "richer societal understanding"? Even the former could well lead to polarized views based on explicit differences of value. Separate technical submissions to a decision-maker (e.g. an Inquiry, or at consultation on proposed draft regulations) would then seem better to me. The balancing of predictions, or "proposed solutions" or other conflicting arguments, based on very different value-systems, must remain a job for the appropriate decision-making institutions of society. They need to consider any presentation, quantitative or qualitative, probing interest, values, status and intellectual rigour.

However the technical risk assessment (or any counter-assessment) is produced, it (or they) should be openly presented, as comprehensible results, in all *subsequent stages of decision*, in which those objecting to a proposal must have proper audience. A series of *current mechanisms* exists, ranging (in the UK) from consultative documents prior to finalization of legislation, to formal legal inquiries and to parliamentary discussion of reports.

The mechanisms are evolving, and the above practices are now more common. They will differ according to circumstance:

- a decision in the field is quite different from the kind of strategic decision discussed here
- a planning matter involving (say) land development near a major chemical plant is essentially different from (say) a decision to license a nuclear plant, and different again from (say) the formulation of a safety regulation, even though all may well involve QRA
- the decision mechanisms may interact with others: e.g. government agency evidence to a local planning authority or an inquiry, or interactions between governments and international organizations
- each mechanism must rest upon the traditions of a country.

There are *many theoretically possible risk discussion forums*. Royal Society (1992: ch. 6) is extremely valuable in discussing the possible range. It does not aim to decide on, but rather critically to list, potential approaches to risk decision processes, and how the various basic risk strategies might affect decisions, and the level of safety achieved.

In discussing "quantificationism and qualitativism", it refers (pp. 160–61) to "important shades of opinion among those who favour . . . quantification" a "rather less influential opposite camp who are uneasy about placing heavy emphasis on QRA", and some "radical critics" who consider QRA actually harmful, because of its tendency to divert attention from non-quantifiable issues. In practice there seems to me no need for opposite camps; rather, a need to take account of QRA with due scepticism, and an attentive ear for the "radical critics".

It cannot be assumed that existing decision processes are either perfect or should be thrown overboard. Thus, the adversarial process generates a losing, and presumably aggrieved, party. It seems much less likely to generate a collective search for improved safety than does the consensual approach; but, in contrast, the latter breeds suspicion of "fixing" from those who do not (or do not choose to) participate, or who have a different world-view from those making the decision.

Nor are other possible processes necessarily ideal. For example, one might question the "collibratory process" that works (ibid.: 167) by "explicitly juxtaposing rival viewpoints in a constant process of dynamic tension with no preset equilibrium" and "on the principle of the desklamp" (see Ch. 9, this volume). Although analogies are dangerous, my desklamp casts light on only a narrow area, which it is incapable of choosing for itself, and is instead determined by an external unseen hand. This does not sound like a good recipe for effective, independent, robust and transparent decision.

Conclusions

QRA is an essential input into a strategic decision process about risk, but its imperfections and omissions need to be borne in mind by the decision-maker. It cannot determine decisions about risk, which are essentially political.

The current decision forums mentioned here allow for this. They are technically informed, as they should be, but cannot be described as technocratic. To this extent, some of the criticisms of QRA are ill founded and they misunderstand the nature of QRA and of current decision processes.

Nonetheless, some real problems remain:

(a) At what point do the complexities and uncertainties of, and known reactions of some sections of the public to, an expert risk estimate become so great as to question its relevance or legitimacy in decision-making?

(b) How effectively can the views of those people who reject the indications of a QRA be taken into account?

(c) In particular, how can assimilation of a rigorously probed but imperfect expert estimate be combined with deep objections that in effect regard expert estimates as irrelevant?

(d) How are (b) and (c) to be achieved without appearing biased to those antipathetic to technological or quantitative ways of thought?

All this is surely the art of government, not the technique of QRA. It seems fair to ask what kind of decision on a technically based risk (and on the underlying technology) might be made without any expert quantitative estimate of risk? It is this that I believe makes it absolutely essential to use QRA, but with all the caution that has been noted.

LIMITS TO THE MATHEMATICAL MODELLING OF DISASTERS

B. Toft

Introduction

In a letter written to The Honourable Mike Gavel of the US Senate, the Comptroller General of the USA reported, in relation to the Senator's concerns over the confidence that could be placed on quantitative reliability predictions, regarding the possibility of a catastrophic nuclear accident that:

> As far as we could learn during this brief review, DOD and NASA officials can offer little guidance as to how very rare failures or catastrophic accidents to systems can be anticipated, avoided or predicted . . . NASA goes to extraordinary lengths – reliability cost is hardly an object – to prevent disasters in manned space vehicles . . . Still, three astronauts were lost in one vehicle. The Soviets suffered similar losses in other attempts. No one can tell if and when such catastrophic failures will be repeated. (Annex to WASH 1400 (NUREG 75/014) October 1975, pp. 197–8)

The deaths referred to by the Comptroller were those of the *Apollo* space capsule crew, who perished in a fire during practice drills in January 1967, and the crew of the *Soyuz XI* space capsule who died following the capsule's decompression during re-entry in June 1971. Since the writing of that letter, other examples have illustrated only too clearly our inability to predict accurately the probability of disaster scenarios occurring. For example, Three Mile Island (28 March 1979), Bhopal (3 December 1984), Chernobyl (26 April 1986), the space shuttle *Challenger* (28 January 1986) and the explosion in Guadalajara, Mexico (22 April 1992) – to mention but a few.

Historically, in the aftermath of such tragic events it has been to the engineering community that society has turned for help in understanding why they occurred and how they might be prevented from recurring in future. Unfortunately, however, the intense public and political pressures generated by the media coverage of these types of events have driven governments the world over, to exhort and encourage engineers to develop methods that appear "objectively" to demonstrate how safe a particular technology, installation, artifact or process is. Thus, since the safety culture of engineers has developed through their particular collection of beliefs, norms, attitudes, roles and practices, it is perhaps hardly surprising that, when confronted with new problems, including those relating to people and management, they tend to use those same engineering techniques, models and assumptions with which they are familiar (Mangham 1979, Pidgeon 1988, Blockley 1991).

Other evidence to support this view is to be found in the ACSNI (1991: 2), where it is noted that, when risk analysts endeavour to measure the probability of human errors:

> Most follow a reliability engineering model, decomposing human tasks into their constituent parts and predicting them in the same way as any other (physical) component in the system.

Furthermore, there appears to be an attempt by some engineers and scientists to persuade decision-makers that prescriptive techniques, of modelling the risks inherent in contemporary technology, represent the way forwards. Of course, it might well be the case that those supporting such a view are simply following the lead of government bodies, who appear to prefer quantitative methodologies, since they can make the regulatory process much easier (Pitblado & Slater 1990). For example, given that some form of methodology for calculating the safety of an installation is available, then a regulatory body can simply direct that any establishment can be granted a licence to operate if it has been predicted that the risk of a disaster meets with a particular numerical standard.

Clearly, having such an apparently "objective" system makes the licensing process much simpler than it would otherwise be, since the assessment of a plant's safety appears to be based upon quantitative objective "facts" as opposed to what might be considered to be unmeasurable subjective speculation. Additionally, the probabilistic numeric derived from such methodologies can be utilized to legitimize claims that the risks associated with a particular hazard are acceptable, since it is possible to demonstrate explicitly that they are numerically smaller than some preset criteria of acceptability and thus can be used to calm any fears that may have arisen within the general public or other interested parties.

Unfortunately, the methodologies used for quantifying the probability that a disaster will occur in any given organization appear to possess at least six significant implicit assumptions, together with a paradox that would seem to render extremely problematic any predictions that might be made using them. As a consequence, making the numerical probabilities derived from such techniques the sole, main or even the partial means of making decisions relating to safety becomes, in many situations, debatable. These arguments will be elaborated in the following sections. Thus, regardless of the underlying convenience that quantitative methodologies might provide, it is undoubtedly true that Waring (1992) was correct in pointing out that "such approaches are usually inappropriate."

Implicit assumptions about risks

The first widely held tacit assumption is that risks can be treated as though they were concrete physical entities that can be precisely defined and unambiguously measured in objective terms. That is to say, risks and their assessment can be considered to be value free and neutral. This notion is, to say the least, contentious, with Otway & Pahner (1980: 157) observing that in individuals:

> The perception of risks is a crucial factor in forming attitudes; obviously people respond to a threatening situation based upon what they perceive it to be.

Douglas & Wildavsky (1982) argued in similar vein that different societies, and the individuals of which they are composed, create their own sets of criteria against which the risks associated with a particular hazardous circumstance will be interpreted and "measured". The use of such social and individual reference schemes suggests that the risks perceived by a given society or individual are not objective but subjective.

This theme has subsequently been elaborated on by several authors. For example, Reid (1992: 151) likewise proposes that:

> . . . it is unrealistic to presume that the fundamental processes of risk assessment are objective.

and Shrader-Frechette (1991: 220) has stated that "All judgments about hazards or risks are value-laden". The most extreme position has been taken by Slovic (1992: 119), who has suggested that "There is no such thing as 'real risk' or 'objective risk'". Drawing upon these assertions, it can be postulated that, to some extent, all risks can be envisaged as being subjective in nature, as was indicated by Pidgeon et al. (1992).

Such an argument clearly undermines the whole notion of an unbiased objective approach to quantitative probabilistic risk assessment and the confidence that society can place on the results derived from such endeavours. For, if the assessment of risks is subjective (i.e. the probability and magnitude of risks only exist in the *mind* of the beholder), then it is not possible for anyone to take objective measurements of a risk as one would of a physical phenomenon. As a consequence, the numerical output of such risk assessment techniques is highly unlikely to produce unambiguous and uncontroversial probabilistic values of a particular risk to which both lay and expert assessors can agree. For example, the pressure group Greenpeace, and those who work in the nuclear power industry, hold diametrically opposed views as to the safety of nuclear technology. However, both claim that it is their particular selection and interpretation of the available data to which society should give credence and, hence, their position with regard to nuclear technology that should be adopted.

Thus, precisely what criteria are to be utilized by decision-makers in determining whether hazards – such as those posed by, for example, genetic, nuclear or chemical technologies – are to be considered acceptable to an organization or society, is a moot point indeed, particularly if the decision to proceed with such an activity is to be based upon a numerical evaluation of the threat. As Gherardi & Turner (1987: 10) have suggested:

> On the one hand we are mesmerised by numbers, even when they are pseudo-numbers, those who deal with them frequently no less than those who are thrown into a panic by them. On the other hand, the general standard of teaching about mathematical issues is so poor that few people understand fully the nature of the properties of the numbers and number systems which they are advocating or excoriating.

Hence, decision-makers would be wise to bear in mind the Funtowicz & Ravetz caveat (1990: 10) that "numerical information is capable of seriously misleading those who use it".

A second assumption is that the modelling of risks is a neutral, objective activity, resulting in a final quantitative assessment that will be unbiased and also independent of the analyst. However, following a recent comprehensive benchmarking exercise into the methodologies available for chemical risk assessment, held at the European Joint Research Centre at Ispra, Italy, and carried out by 11 specialist teams representing a wide range of interests, including regulatory bodies and industries, Amendola et al. (1992: 355) reported that ". . . the numerical results are strongly dependent on the assumptions adopted . . ." and that

> When the results from a fault tree analysis were compared with historical data used by the other teams for the same event, substantial differences were found. This could be due both to the assumptions made and to the data adopted for the primary events.

The problem of bias is also highlighted by Watson & Buede (1987: 280), in a discussion of decision analysis (and quantitative probabilistic risk assessments are often one of the inputs to that process), when they note that:

> . . . once into an analysis, it might be the sponsor, a public official, who will insist that the analysis is carried out in a certain way, perhaps to support a particular agreement.

One example of this type of behaviour is cited by Freudenburg (1988: 243) who quotes that NASA:

> . . . pressured one consultant to produce a more optimistic estimate of booster safety and disregarded even more pessimistic predictions contained in two subsequent studies.

The fact that quantitative probabilistic risk assessments are open to this type of pressure is further support for the view that it can be dangerous to employ the output of such studies.

A third assumption, implicit within the methodologies of quantitative probabilistic risk assessment, is that it is possible for the person(s) undertaking a risk analysis of an organization to specify, unambiguously, an *exhaustive* set of failure modes for the activity, or activities, under consideration. However, systems theory predicts that any open system, in this case an organization, can arrive at a given end state from different starting conditions and via different routes (von Bertalanffy 1968). Thus, it can be argued, that there are an *infinite* number of equally likely ways by which an organization (or a person) can arrive at, or be responsible for, an accident (Lewis et al. 1978, Elms & Turkstra 1992).

The implication is that when one finite set of failure scenarios has been calculated for risk analysis purposes, there is another equally likely set waiting to be calculated, and so on, ad infinitum. Therefore, the risk probabilities that are calculated regarding a finite number of failure modes are in one sense meaningless, as there are always other, equally likely, ways in which an organization, or operation, can meet with a disaster which have not been considered. One empirical example that supports this view, is to be found in the 1983 Royal Society report into *Risk assessment*, where it is pointed out (Royal Society 1983: 192) that:

> Charles Komanoff, reporting on the Three Mile incident, says that the sequence of events which caused the accident at Three Mile Island was not among the supposedly exhaustive list of possible initiating chains for reactor accidents in the Rasmussen Report.

A fourth supposition is that reliable historical data is available for past events that can be utilized to calculate the risk probabilities. With regard to this assumption, the Royal Society's 1983 Report makes the point that data on the frequency of unwanted events is in short supply, and that, because the events are rare, the techniques used for sampling do not provide sufficient data. Indeed, with regard to data on physical devices it was reported at the Piper Alpha Inquiry (Cullen 1990: 307) that:

For subsea valves Dr. Gilbert stated . . . He doubted if data on the probability of successful operation on demand of any large population of such valves were available.

Without reliable high-quality historical data to hand, it is hard to envisage how a realistic, quantitative, probabilistic appraisal can be carried out, since the level of uncertainty regarding the validity of any conclusions will be extremely high (Collingridge 1980, Pidgeon 1988, Funtowicz & Ravetz 1990, ACSNI 1991). Lewis et al. (1978: xi) advise that risk analysts should:

In general, avoid use of the probabilistic risk analysis methodology for the determination of absolute risk probabilities for subsystems unless an adequate database exists and it is possible to quantify the uncertainties.

Yet another presumption is that the complexity of human behaviour in general and human errors in particular can be pre-specified and reduced to a simple unitary numerical representation. So far as can be ascertained, there is no publicly available model that can accurately forecast human behaviour, let alone predict the probability and types of errors that different people might commit under diverse, and in some cases extremely dangerous, circumstances (Grose 1987, Reason 1990, Elms & Turkstra 1992).

The ACSNI Report (1991: 2) also notes that:

The simple method of multiplying probabilities together assumes that each is independent of the others, which is true in many physical engineering situations; but clearly often untrue in human ones.

These observations again raise serious questions as to the legitimacy of a numerical expression derived from any of the currently used methods for forecasting human behaviour.

Finally, there is an implied assumption that the future trajectory of an organization will be similar to that of the past. However, an organization is a dynamic entity that is constantly changing and, although the past cannot be altered, the future can. As a consequence, any forecasted numerical probabilities of a risk eventuating would appear to be of doubtful relevance, since any future unpredicted misunderstanding or action by an employee can change the calculated risk factor of a major disaster from one in a million to an absolute certainty.

The open systems paradox

The paradox to be found in the generation of quantitative probabilistic predictions of risk to any organization, stems from the notion that in any system the whole is always greater than the sum of its parts. For example, a bicycle when stripped down into its components and placed in a box is nothing but a collection of engineered steel, rubber and plastic parts. However, when assembled in a particular way, the synergy created transforms those parts into the elements of a system from which a property emerges that the separate individual components did not possess. That is, when the parts are combined to form a bicycle, the output of the assembled system can be used as a mode of transport. Thus, the assembled bicycle has *greater* utility than the sum of all its dismantled constituent parts. This same argument can be employed to characterize the properties of an organization, that is, an organization is greater than the sum of its constituent parts.

However, when a quantitative probabilistic risk assessment is carried out on an organization, a finite number of failure modes for each hypothesized scenario is derived, the risks calculated and then summed together to produce a final risk probability prediction. Therefore, the final calculated risk of a system failing in a catastrophic manner is exactly *equal* to the sum of its finite hypothesized parts. It is the use of the methodology that creates the paradox; for a system, in this case an organization, cannot be greater than the sum of its parts and at the same time be exactly equal to them.

The paradox is created because organizations are, as noted above, open systems. However, quantitative probabilistic risk analysis methodologies were originally, and still are, used to model the risks associated with manufactured artifacts, that is, closed systems. Hence, paradoxically, an attempt is being made to model and calculate the risks in *open systems* using *closed systems* techniques.

Consequently, it can be argued that the techniques of quantitative probabilistic risk analysis are inappropriate for the evaluation of many of the risks to which organizations are exposed and are, therefore, of limited usefulness.

Examples of failures

Several empirical examples can be used to illustrate the futility of trying to calculate numerical probability values in such open systems. The first concerns the unexpected human behaviour linked with the Ekofisk blowout on 22 April 1977, where the fail-safe equipment was actually overridden by the operators on duty (Royal Society 1983: 192). Another example is the behaviour of the engineers at the Chernobyl atomic power station in removing the safety interlocks so that they could proceed with their unauthorized experiments on the reactor. A third, more recent, example, is the Japanese engineer who inadvertently left an aluminium rod inside a console at the Okuma nuclear power plant in October 1992 after carrying out fault-finding tests. The rod caused a short circuit, which led to an error signal being generated, which caused the nuclear reactor's computer incorrectly to close down all three high-pressure water pumps belonging to the core's cooling system. This action, in turn, led to the water covering the reactor's core to drop by around 900 mm and to the triggering of automatic safety systems that quickly shut the reactor down (Hadfield 1992).

It is doubtful if such scenarios would have ever been contemplated in the first place and, if they were, clearly the numerical probabilities generated would have made each of the incidents appear so unlikely a course of events that no action would have been taken to prevent such behaviour from occurring.

However, this is not to argue that the quantitative analysis of the properties of manufactured closed systems and their components have not made major contributions to safety, for they have. Indeed, the HSE (1989b) Royal Society (1983) and ACSNI (1991) reports all take the view that such quantitative analyses are useful but that they must be used carefully, with discretion, and in the *appropriate* circumstances. For as LaPiere (1934: 237) observed:

Quantitative measurements are quantitatively accurate: qualitative evaluations are subject to the errors of human judgement. Yet it would seem far more worthwhile to make a shrewd guess regarding that which is essential than to accurately measure that which is likely to prove quite irrelevant.

The ACSNI (1991: 20) report in particular recognizes this problem, for in the conclusions drawn with respect to the techniques used in human risk assessments, it is noted that:

... there is benefit if the analyst first carries out a systematic and comprehensive *qualitative* analysis of possible operator errors. [present author's emphasis]

Additionally, Karl Popper (cited in Horgan 1992) argued that:

Determinism means that if you have sufficient knowledge of chemistry and physics you can predict what Mozart will write tomorrow ... Now this is a ridiculous hypothesis.

Thus, it would appear that the utilization of prescriptive and deterministic quantitative models for the assessment of risks in organizational settings is not unproblematic (Ansell & Wharton 1992).

Finally, quantitative analysis, so much a part of the engineer's and natural scientist's culture, may be so deeply inculcated into the fabric of their existence that it might be acting as a mental barrier, similar to the "groupthink" phenomenon (i.e. a socially derived form of "mind set") identified by Janis (1971). And, as a result, it could be preventing the exploration of other, potentially more robust, management orientated approaches to the assessment and evaluation of risks, particularly, as many of those who sit in powerful positions regarding the distribution of resources come from such backgrounds.

The future

One way forwards in the search to overcome these difficulties might be the use and further development of the notational scheme devised by Funtowicz & Ravetz (1986) to describe the expression of quantitative information. This system of notation is designed to characterize explicitly the way in which a number has been derived. When applied to the product of a calculation, it provides a decision-maker with an indication as to the amount of reliance that can be placed upon the accuracy of a number and hence, whether or not the number has any "real" contribution to make to the activity under consideration.

Another approach might be to have an organization at a national level dedicated to the collection, analysis and dissemination of information gained from all types of organizational failures (Toft & Reynolds 1994). Like any other organization or system, such a body should not be thought of as a single level structure, but as incorporating linkages between several levels and a variety of organizational learning process. Figure 4.1 is one way of depicting such a structure.

The model illustrated is an attempt to create a theoretical "system of systems" which might help to reduce the number of organizational failures. It is sketched out here not as an immediate policy proposal, but in order to highlight those issues which any such system would need to confront. In the model, the terms "environment", "design" and "management" should be interpreted very broadly, so as to include all those situations and personnel involved in the development and operation of a sociotechnical product, organization or procedure.

Referring to Figure 4.1, a sequence of events would typically begin with someone perceiving a need for change in the "environment", such as a new bridge, aircraft or service. Once the need has been recognized,

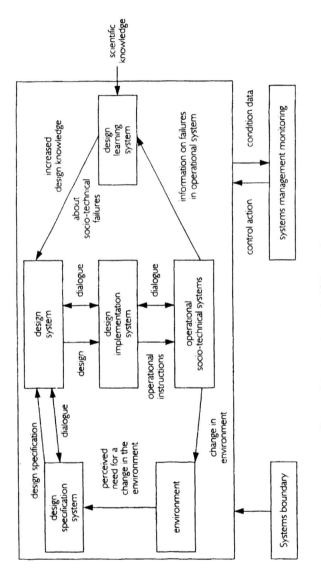

Figure 4.1 A sociotechnical failure reducing system.

and sufficient financial backing made available to support the project, the next event would be specification of the precise attributes of the finished product, organization or procedure. This would occur within the "design specification system". For example, the specification might be that a bridge should be capable of withstanding a load of "x tonnes", or that an aircraft should be capable of carrying "y passengers".

Having decided upon its parameters, the design specification would now be submitted to the individuals and organizations contributing to the "design system". Here the specification would be translated into a series of working plans representing the finished product. After the design process is completed, and the design accepted by the client, it would then be transferred to the "design implementation system". At this stage the design would be costed, contracts and subcontracts issued, and the construction or building work undertaken. When the implementation process had been completed, operational instructions would be issued. The "operational sociotechnical system" so created then begins to act, and in so doing changes the environment in the manner originally stipulated in the design.

What if the sociotechnical system now fails – say, a bridge collapses – or the change in the environment creates unforeseen problems for the other organisms sharing that environment? Among the elements of the model described so far, there is no mechanism for the lessons from all such experiences to reach every one of the sections of society that might usefully gain from such information.

The right-hand loop in Figure 4.1 is such a mechanism – a "design learning system" – that would enable the lessons drawn from failures to be incorporated into the working practices of the future. To achieve this, the "design learning system" would need to collect, collate and analyze data from known organizational failures, and also incorporate knowledge gained in the "sciences", thus fusing together both practical experience and academic research. As a result, new statistical techniques to predict failures might be developed and theoretical predictions compared to empirical findings (Strutt & Allsopp 1993). Innovative ways of managing safety could be devised and tested against known empirical problems. This output would be valuable knowledge for both the design and management functions.

The knowledge would be particularly valuable if it could then be made to impact upon the designers of organizations as well as on the legislators and managers who control such systems. Beyond this, information relevant to the management of organizations could be communicated directly to the industries concerned, to educational establishments and professional institutions, and to other interested parties. The whole system of systems could perhaps be managed and monitored through the creation of a multidisciplinary professional institution, where both social and physical sciences would be seen as equal partners (Freudenburg 1988), or through the extension of the activities of a government body, such as the HSE which, although it does have a wide remit at the present time, does not cover all types of technologically based disasters.

Conclusion

In conclusion, it can be argued that not all the problems noted above are intractable. If the "notional scheme for the expression of quantitative technical information" devised by Funtowicz & Ravetz, and the hypothetical design learning system described above, were to be implemented, some of the present limitations would be addressed and in time that might lead to improvements in the ability of quantified probabilistic risk analysis to model organizational risks.

However, at the present time the evidence strongly suggests that the methodologies currently in use to calculate the probability of organizational disasters occurring suffer from empirical and theoretical limitations. Currently, these limitations are so profound that to employ such calculations in the criteria for taking decisions regarding the acceptability of many classes of organizational and societal risks is clearly fraught with danger. Miller (1962: 95) made the point succinctly when he observed:

> In truth, a good case could be made that if your knowledge is meagre and unsatisfactory, the last thing in the world you should do is to make measurements. The chance is negligible that you will measure the right thing accidentally.

CHAPTER FIVE

Designing institutions: a house of cards?

THE FEASIBILITY OF INSTITUTIONAL DESIGN IN RISK MANAGEMENT

A fourth key element in contemporary risk management debates (more often implicit than fully articulated) turns on whether or not there is a reliable knowledge base on which to ground effective institutional design for risk management, and the extent to which the orthodox engineering approach to "design" can feasibly be extended to complex organizational structures, especially in social and sociotechnical systems. The majority of recommendations for change which come in the aftermath of major events/incidents/accidents, tend to be social and administrative, but it is in this very area (especially where complex systems are involved) where the greatest doubts arise about the reliability of the knowledge base on which institutional design can be built.

At one end of this spectrum are those who argue that sufficient understanding exists about how institutional design affects vulnerability to system failure and risk-taking behaviour for principles of good practice to be articulated with some confidence. For example, Breyer (1993) argues – perhaps against the temper of the times – that better-developed bureaucratic expertise along the lines of the French *Conseil d'État* is a better prospect for developing an approach to risk management that is authoritative, rationalized and insulated from "random agenda selection" than "recently fashionable" alternatives such as reliance on tort law or deregulation. Research in this area includes work on organizational vulnerability to major system failure (Horlick-Jones 1990, 1991, Horlick-Jones & Peters 1990, Horlick-Jones et al. 1991); work on "high-reliability organizations" (Halpern 1989, Roberts 1989, Roberts & Gargano 1989, Rochlin 1989, Weick 1989; Sagan 1993: 14–28 and *passim*); and the work of Reason (1990) and his colleagues on human error. Those who see considerable scope for the development of established principles of institutional design point to what they see as the cumulative growth of a recognized practice of corporate safety management since the mid-1980s, particularly in multinational corporations that have become concerned about the growing visibility and rising costs of large-scale accidents.

111

Spearheaded by some of the major corporations in the chemical and processing industries, such practices tend to stress a commitment to safety at the highest organizational level, the adoption of low- or zero-accident targets on a corporation-wide basis, and the provision of training and resources to back up such attention. Corporate safety management programmes include regular safety audits, the inclusion of more safety elements in total quality management programmes, and the promotion of arrangements for continuous organizational learning such as quality circles and "toolbox safety meetings", in order to raise safety consciousness in all parts of the corporation. The standard of a corporation's "safety culture" has also occasioned considerable discussion following the introduction of this term as part of the assessment of the after-effects of the 1986 Chernobyl nuclear power station meltdown (cf. CBI 1990).

At the other end of the spectrum are those who are much less optimistic about the scope for, or the knowledge base underlying, "institutional design" claims in risk management, especially in large sociotechnical systems. These sceptics point to major limitations in the current state of knowledge about how risk is handled in human organizations, and consider there is a much less robust knowledge base on which to ground design than exists in traditional engineering. However, assumptions about organizational design become increasingly crucial as the development of technology comes to involve constructing ever more complex systems and changing the nature of human/machine interfaces. For example, some software specialists are sceptical about the extent to which software development is part of the solution or part of the problem in engineering for safety, given that design integrity in this case requires the capacity to anticipate and cope with interactions between complex and highly volatile technology and the dynamics of human response to that technology. In addition, some authors are sceptical of claims that organizational "safety cultures" can be deliberately engineered by management or that such doctrines are grounded in systematic investigation of cases rather than selective anecdotes and analogies. Such sceptics note that in practice there are considerable variations in corporate safety policies in transnational corporations, and sometimes even apparently contradictory approaches seem to yield similar outcomes (as in the cases referred to in Ch. 3 concerning liability).

Those who favour an emphasis on institutional design in risk management are entitled to question whether it is helpful merely to bring scepticism to the risk management debate; the sceptics have sharp questions to ask as to how much straw the designists really have in their bricks. The growing volume of research in areas such as human/computer interfaces and user-orientated architecture. ought to enable those who wish to expand the scope of institutional design to make some ground;

and, in principle, a constructive dialogue between the two approaches ought to guide experimentation and to enable each to develop its position well beyond first principles.

In the next section of this chapter, David Weir approaches his own particular brand of "designism" via the argument that planes crash and businesses fail for essentially similar reasons. Then Edmund Penning-Rowsell argues that, despite the accumulation of considerable knowledge on hazard and risk, particularly in the field of natural hazard research, an extremely poor level of understanding exists as to how to design institutions to cope with the complex realities of risk management.

RISK AND DISASTER
The role of communications breakdown in plane crashes and business failure

David T. H. Weir

Systems, failure and guilt[1]

On the whole, things don't work. Sociotechnical systems will fail, even if they are designed to minimize the possibility of human intervention and are structured by a hard technology which may be very advanced. However, much of the teaching and research in the field of complex systems ignores, or skates over, this undeniable truism.

The paradigms of engineering and the social sciences alike are predicated on assumptions of the perfection of hard-wired systems and of the perfectibility of human systems. Their language is replete with terminology indicating confidence in these phenomena. We persist in the belief that systems are "robust" and markets are "efficient", and that we learn from experience. But the one thing that experience teaches is that it does not perfectly inform.

In 1970, I moved from the solid theoretical certainties of a sociology department to the turbulent eclecticism of a rapidly growing business school. My special claims to academic competence were rooted in criminology and organizational theory. In considering what these two branches of knowledge entitled me to teach experienced managers, I combined the insights from both to create a new programme entitled "Trouble at t'mill; the theory and practice of organizational deviance". This was, to my initial surprise, an immediate success as a course, because it reflected the experience of reality of most of the participants who were experienced senior managers. They knew, in their collective water, that however brilliant and clear-cut the plan, there would always be tears before bedtime.

The only surprise was that such a taught course did not already exist. Attempts in other business schools to introduce a similar approach are often met with the polite but dismissive judgement of those teaching the "mainstream" topics of economics, marketing, finance and business policy, that these themes of deviance and organizational failure were indeed interesting but marginal.

---

1. Earlier versions of this section were given as papers at the ESRC Seminar on "Systems failure, hazard management and industrial design" at the London School of Economics and Political Science and the Work Organisation Research Unit, University of Bradford Management Centre. The helpful contribution of these colleagues is acknowledged.

It requires a sociology of knowledge to analyze the reasons for a culture's determination to reject the blindingly obvious and it is not proposed to attempt that here. The study of organizational disaster is a worthwhile one, though, because the phenomenon is relatively common-place and still insufficiently understood: but questions remain as to the most appropriate methodology and sources of theoretical paradigms upon which to found the research enterprise in this field.

A suitable framework for the analysis of failure in complex sociotech-nical systems may lead to understanding of the common elements in the aetiology of apparently dissimilar disasters and to create a basis for future research. These factors may be sought in the area of organizational com-munication. The material exists already for a preliminary taxonomic and typological analysis. This would clarify the nature of communication in complex organizations and assist in the identification of probable sources of vulnerability in sociotechnical systems, thus enabling the more precise targeting of research into these phenomena.

Much research has concentrated on overt disasters, such as major fail-ures in transport systems typified by the Zeebrugge ferry sinking, the King's Cross underground station fire, the Clapham and Purley train wrecks. Other disaster analyses have concentrated on major crashes and fires involving passenger aircraft, such as the British Midland (Kegworth) crash and the Tenerife collision. Other writers have reported failures in complex structures and installations, such as the Flixborough explosion and the *Challenger* space shuttle catastrophe.

One initial motivation for concentrating on events of this kind lies in the loss of life. This engages human sympathy and has sociopolitical implications. Often these involve the perceived necessity to assign blame rather than to achieve understanding. An analysis of "cause" may be the precursor to a finding of "guilt" (see pp. 72–83).

Douglas and others have shown how widely situated are these modal-ities in human cultural systems (Douglas 1992). They are, by no means, merely found among the apparently erroneous explanations of primitive peoples (Evans-Pritchard 1932).

Other imperatives for the study have come from the need to ameliorate the consequences of disasters (Drabek 1986). Studies of emergency plan-ning and community reactions are concerned with administrative effi-ciency, either as a prophylactic against future disaster, or as a mechanism for assuaging the collective guilt of a community (Quarantelli 1978). But a system, however well conceived, for ordering and clarifying the admin-istrative consequences of disaster, cannot of itself prevent these events happening. Indeed, too close attention to the initiating causes of any single disastrous event may exacerbate the damaging consequences of subsequent events, even those of essentially the same kind.

It is sufficient for the moment to say categorically that disasters will happen, do happen, and always have happened, although not in an obviously predictable fashion. There is an essential randomness and indeterminacy about this kind of event which is intrinsic to our conception of it as a "disaster" in the first place.

There are various phases in the evolution of a complex system. Before anything else, there is a period of system design leading to the inception of the system, followed by one of partial trial, before the system enters the stage of what is usually described as "normal running". In this period of apparent systemic normality, behaviours and outcomes are believed to be predictable on the basis of the managers' understanding of the design of the system itself and the environmental conditions it encounters. Where mistakes, problems, or minor events requiring amendment of the initial system occur, as they inevitably will, then overt improvement or "patching" of the system may be introduced, and will become integrated into the normal operation of the improved system subsequently.

But systems in general do not follow an even life-cycle from perfection to failure. *All complex sociotechnical systems tend to operate in degraded mode.* Under normal operating conditions, the actual state of the system will usually contain improvements, short-cuts, error-correcting routines and other elements "patched" into the system in response to local failures, or on the basis of feedback about the conditions actually experienced in operation. Many of these "patches" will not be documented.

Most complex systems operate in a sociopolitical context externally and internally, which distorts the information basis on which system performance is constructed in terms of which system activity is described. *It is thus normal for the system to be "abnormal" in its operation.*

So, to the language of system, feedback, inputs and outputs, must be melded the fundamental sociological categories of power, domination and intention. Of course, systems vary in the degree to which the sociopolitical values of the wider environment interpenetrate with the control structures of the system itself. But it is possible to identify systems in which the *demand* for control is so great that the *achievement* of control becomes more problematic.

Consider the example of theft within a prison. A prison is a system of social control relatively insulated from the host environment except at key points (Morris 1963). Its purpose is to create conditions for the punishment or rehabilitation of the individuals incarcerated. It is a people-processing institution. It normally operates within the framework of an overtly articulated legal system in which it occupies an important symbolic position. Thus, the purpose of a prison is to encourage behaviour that conforms with the requirements of both criminal law and social morality.

But the most minimal acquaintance with prisons indicates that the reverse tends to be the more normal condition. The structure of rules is *so* closely articulated, and its relation to social control *so* tight, that it is almost impossible to achieve control of even quite simple behaviours, except through the undertaking of illegal acts. Roles, positions and responsibilities are so tightly prescribed that the most plausible method of obtaining articles in short supply is to induce the commission of a further crime. The prisoners, and especially the least trustworthy, most dangerous, and least malleable among them (often called "trusties"), have more degrees of freedom and thus more control over their life and that of the officers, than those who make efforts to conform to the moral and legal expectations of the prison.

Many bureaucracies, of course, exhibit the same paradox and many sociologists have documented this form of degradation. But it is a feature of all imperatively coordinated associations.

To Bertie Wooster is attributed the memorable remark that "The problem with truisms is that, dash it, most of them are true". The most widely quoted truism in management culture is that embodied in Murphy's Law, which, in its simplest form, states "If a thing can go wrong, it will" or, more graphically "The toast will always hit the floor butter side down". The fact that these aphorisms provoke a wry smile of recognition does not render them useless as indicators of the endemic phenomena that all managers recognize.

Of course, it is equally correct to say that, on the whole, most systems do work, one way or another. But they tend to operate in degraded mode. It is also probable that, in most cases, the extent of the degradation is not fully understood, because it is only partially documented, if at all.

There are various reasons for this. The reporting systems and document production routines within an organization are normally related to the predicted operation of the system under perfect conditions. By definition, accidents and random interventions of a negative kind are not predicted, and thus may not be allowed for in the formal systems of reporting. When events occur that go contrary to the expectations built into the system, there is a tendency to disregard them, or, in extreme cases, to believe that the event has not happened at all. Those who design complex systems become emotionally and politically committed to the view that the systems are working along the intended lines. Those in positions of power and authority support the belief that the systems on the whole are good, achieve the ends intended and operate on predicted lines.

So, evidence about failure is normally regarded as at least subversive, if not overtly damaging. Sometimes this disbelief is expressed in such terms as "confidence". Thus, to raise doubts about a system's operation is criticized because it will "damage confidence" in the system's efficiency. A "subterranean" culture can be created in which it is possible to achieve understanding of how the system is actually working only by reference to information sources that are outlawed by those in control of the system (Matza & Sykes 1961).

Bensman & Gerver (1963) show that these relationships between the formal and the informal systems may be very complex and sometimes counter-intuitive. In the case of the illicit operating procedures known as "the tap", which operated in an aircraft manufacturing plant, it was the most skilled and most trusted operatives who were charged with the responsibility for undertaking actions explicitly outlawed by the formal system. Only they had the skill to conceal their illicit practices.

A further reason lies in the general human demand and desire for order. When things appear not to be working in understandable and pre-dictable ways, there is a danger of organizational and personal demoral-ization. People need to believe that things are, on the whole, working out, even well beyond the point where empirical evidence would indicate the contrary. Perceptions of systems tend towards a closure of the perceived explanatory structure in ways that reinforce supportive patterns of belief and, in particular, the belief that someone, even if not the particular actor is, or could be, in control.

Managers who are highly trained, especially in technical areas, believe that their own training gives them powers of control; they thus interpret phenomena in terms of the pattern of explanation for which they have been trained. Sometimes this involves the neglect or the negation of evidential phenomena that would have a different meaning for managers trained in another pattern of explanation. Often it is the untrained or naïve, or low-status observer lacking the emotional commitment to a particular type of explanation or not seeing the need to protect a political or organizational decision, who most clearly perceives the impending danger.

Communication failures in catastrophe

Failure in the communication system, and in particular in the recognition and transmission of potentially catastrophic conditions, is evident in many major disasters.

Thus, in the British Midland 737 crash on the M1 at Kegworth (1989), the pilots shut down the right-hand engine, which was operating nor-mally. However, a few seconds earlier the passengers and cabin staff had seen flames, experienced vibrations and smelt smoke from the left-hand engine (Grayson 1989). Those in the passenger cabin did not tell the cap-tain what they saw and heard. Had the captain been told, he might have recognized his mistake, restarted the good engine and shut down the defective one. But it is also significant to understand the social and organ-izational mechanisms that inhibited the passengers from advising the captain of his error. This inhibition extended even to the cabin crew, who, in the words of the chief steward, "thought he [the captain] must have known already because everybody could see it". But had the information been passed to the captain from these low-level informants, he might still have preferred the apparent evidence of his instruments, his experience and his judgement, and taken no notice anyway.

Communication factors are equally crucial in the analysis of the *Challenger* space shuttle explosion in 1986. The immediate cause of the disaster was a blow-by of superheated gas past an "O" ring and putty seal in the rocket motor. But engineers from the Morton Thiokol Company, including the supervisor and senior scientist concerned with the seal task-force, had advised strenuously for a postponement the night before the launch. Their anxieties as to the possibility of a catastrophic failure were precisely borne out in the subsequent explosion (Institute of Electrical and Electronic Engineers 1987).

It is tempting to ascribe blame and to cast the responsibility for the decision to launch on an uncaring and callous management. But a detailed analysis of the testimony of the Morton Thiokol Company engineers and the launch management to the President's Commission Inquiry reveals something much more subtle and of more widespread importance. *Engineers and managers simply do not speak the same language*; they do not share the same frame of reference.[2]

Hierarchy and frame of reference

Studies of organizational communication illuminate two characteristically problematic situations. In rigidly hierarchical systems, there are overt barriers to the free flow of information, even when that information is of a kind that is crucial for effective managerial decision-making. Another type of failure of organizational communication occurs when there are strong vertical divisions, as between professional groupings who use a specialist jargon and have strong differences of status or formation between them. In the *Challenger* disaster, these two types of organizational communication failures overlapped.

Other problems are commonly found in the analysis of complex disasters. Organizational power, and its operation in a situation of complex interdependence between many institutions and agencies, creates communication problems. Although the individual hierarchies of the several organizations involved provide clear and predictable structures of power and authority, there are problems at the interfaces, and in the grey areas

2. Intercontinental Aviation Safety Consultant's Report: quoted in *Fortune* magazine, February 1989.

where the remit of power structures is not clearly enough specified.

The engineers in the Morton Thiokol caucus clearly felt that, in engineering terms, they should have had, some right of veto over the proposed launch. But this was not clearly specified, the rules were not articulated, and the appropriate behaviour was not triggered. Each of them separately felt that they had been overruled and did not undertake the behaviour that a trade union caucus might have used to lead to collective action and to the operation of an effective veto. They were, in the event, like non-executive directors in the boardroom of a company that is starting to collapse.

Problems of authority, and the locus of it, and of power, and the operation of it, are all commonly neglected in the study of disaster. It is generally understood that "human factors" play an important role in major transport disasters, but it is not always recognized how widespread the phenomenon is. A recent analysis of airline accidents concluded that 67 per cent of fatal accidents involved human error[3] and a further seven per cent involved defects or failures of the maintenance system. Eleven per cent were attributable to sabotage and military action, and only 15 per cent to exogenous factors not immediately related to the sociotechnical system.

Instrument failure and on-board social systems

There is a complex relationship between on-board and environmental features of the situation confronting the management of large jet planes in motion. The flight deck crew are involved in a complex series of communication interfaces. They must refer regularly to their own instruments, monitor, control and adjust their position relative to observations, set and reset for altitude, control, speed, thrust, fuel utilization, attitude and so on. They must be in regular touch with a series of controllers who successively handle the plane in the various stages of its flight from pre-take-off, through take-off to climbing to level flight, to descent and landing. They interface directly with other members of the crew who do not have flight responsibilities, and also indirectly (but sometimes directly) with the passengers.

The on-board dynamics among flight crews have often been studied. United Airlines in particular has paid special attention to the interaction between team behaviour, personality and task. A feature of some accidents has been a lack of communication between very senior, experienced and sometimes over-confident captains and first officers, and second officers who lack those qualities. In some cases, United Airlines

---

3. But see Johnston, this volume, pp. 72–83.

studies indicated the existence of a process in which the more experienced and time-served members of the crew, whether pilots or engineers, banded together against the less experienced.

The hierarchy of authority on board is quite rigid. The captain is, in emergency, in full control and ultimately responsible for safety. It is, therefore, important to consider to what extent there may be structural and behavioural impediments to the captain becoming as fully informed as possible. He or she should be the best informed actor in the situation, but this may well not be the case.

The behaviour evidenced in some reported incidents approaches close to what Thompson & Wildavsky (1986) refer to as "the paradigm protection behaviour" which occurs in hierarchies. The elite status of flight crew, their high sense of professionalism and camaraderie, the rigid hierarchies and quasi-military uniforms – all combine to produce the behaviour of which George Bernard Shaw warned when he described professions as "conspiracies against the laity". In the British Midland M1 crash, the real situation was quite visible to the passengers and cabin staff, but the communication barriers filtered the critical information out of the system. In the extreme situation evidenced in a United Airlines crash at Chicago O'Hare Airport in 1968, the captain almost wilfully refused to accept corrective information from his co-pilot and indeed conspired with the engineer to put down the co-pilot as a valid source of any information. Sometimes, the captain becomes immersed and lost in a communication world of his own. In a crash in Bali in 1982, in which a plane descended too soon after the inadvertent misreading of a flickering compass needle, the captain and flight crew suspected the accuracy of their instrument-derived position, but did not act on their doubts. An Eastern Airlines crash in the Everglades in 1983 exhibited a similar sequence of events. A well known case is the Palm 90 Air Florida take-off crash at National Airport, Washington, on 13 January 1982, where the co-pilot said as the heavily iced plane started its take-off roll, "Gee, that doesn't look right . . .". But he made no explicit attempt to abort the fatal take-off manoeuvre, despite his concern (MacPherson 1984). This concern for the etiquette of hierarchy had fatal consequences for him and for a planeload of passengers and crew. When a Pacific Southwest 727 overran a light aircraft in conditions of perfect visibility in 1978, the crew presumed they had passed the plane, though they had no positive evidence to confirm their presumption.

Many of the so-called "instrument failure" or "instrument misreading" failure events are, in fact, much more complex in their aetiology than this, apparently clear, definition implies. It is the interface of the instrument with the other aspects of the complex sociotechnical system that conspire to cause the catastrophe, rather than any simple mechanical or physical failure in the instruments themselves. Much attention is subsequently placed in inquiries and analyses on the design of the instruments, their visibility and location in the cockpit. But although these are important features and improvements can always be made, they are not by any means as significant as the way the sociotechnical systems themselves operate.

Three different kinds of effects can be distinguished in these cases (Table 5.1). Some are associated with the *hierarchical structure* of the organization. Others are associated with the differing *frames of reference* brought to the situation by members from different professional and technical groups. Still others are indicative of a collective refusal by crew members to accept the evidence of incipient disaster, because of what Janis (1972) called "groupthink".

Table 5.1   Types of communication failure in big plane disasters.

	Example
Authority barriers e.g. co-pilot diffidence, disregard of cabin as source of information	British Midland 737, Kegworth, 1989
Frame of reference e.g. belief that expected message has occurred	KLM/PAN-AM 747, Tenerife, 1977
Groupthink	San Diego/Pacific Southwest 727, 1978

Some incidents are triggered by the belief of the captain, or the flight crew together, that the plane can be flown safely in a hands-on mode in cases of emergency. The reality is that most complex systems, including computer-controlled jet airliners, cannot be managed except on a *team* basis. As a consequence, some airlines have started positively to de-select pilots on the basis of what can be termed the "Biggles factor". The most technically competent individuals in terms of motor skills, visual acuity and personal confidence, may be also those whose reliance on these very same abilities could, in a critical situation, bring about the catastrophic event. It is the team player, not the individual hero, who can save the craft in these situations.

A common sequence of events is that an initial technical malfunction creates evidence that is visible on instruments. Corrective action is taken with reference to the presumed state of the system. However, this may be based on an erroneous judgement, so a secondary cycle of failure is initiated. This gives rise to evidence that is potentially available, either directly or through instruments. But, because it does not fit the frame of reference created by the judgements about the original sequence of events, this secondary information is suppressed. The result, in due course, is a catastrophe. Often the catastrophic event is more serious than would have been the case had *no* action or limited action been taken on the basis of the original evidence.

Communication factors in business failure

The same is true of business. It is very often customers and clients who see the first signs of impending business failure. The hierarchy's frames of reference and reliance on individual skills, which are widespread in the ranks of senior businessmen, make the sort of evidence that comes from clients and customers unavailable in a timely enough form for it to be of positive value for correction, by management, of the pre-catastrophic sequence of business decline.

Over 70 per cent of airline accidents involve human agency, and the bulk of these involve communication failure. Business failure shows a similar pattern. Inadequate management is involved in 85 per cent of cases of corporate failure and 73 per cent are connected with failures of senior management. This compares with 7 per cent, for example, of cases caused by exogenous changes in the pattern of demand for a company's products. Of course, these are not necessarily mutually exclusive categories, but most of the studies of business failure conclude that it is *internal* factors rooted in the management system, and its inability to recognize signs of impending problems before they become catastrophic failure, that are most worthy of attention. Some of the main management problems are reviewed by such writers as Argenti (1976), Slatter (1984) and Kharbanda & Stallworthy (1986). "Lack of communication" is a consistent theme.

There may be considerable similarity between the big jet crash and the company failure in terms of the breakdown of communication systems. The linkages between management and its own sources of information, its key ratios, its management accounts and its visible measures, are strong. They are connected with the day-to-day operation of systems of which management believes itself to be the master. Management may know too much about this kind of information to be able to "see the wood for the trees". They are on the flight deck, insulated from their passengers.

In terms of the external systems, the linkages with the institutions, the banks and the City, are legally strong and intermittently powerful, but the utilization of them as *sources of information* is more or less at the discretion of management. Catastrophic business failure involving bankruptcy could often have been avoided had the senior management informed their bankers earlier of the *actual* state of affairs. But it seems to be normal for senior management to underestimate the nature of the crisis and overestimate its ability to manage it.

In plane crashes, external sources of information are not brought into the picture early enough and at a time when their information could help the management of the on-board crisis. Likewise, the role of such institutions as the Midland Bank, the DTI and the Bank of England in business failures such as the collapse of Barlow Clowes is worth examination.

What the "on-board" management often fails to realize is that it does not control the whole system. The players in the larger game can also talk among themselves.

An extract from the final eight minutes of flight recording of an Eastern Airline jet approaching Miami with a complete three-engine flame out indicates this. From the intervention recorded at 13.36.15 h to final landing at 13.44.58 h, the centre logged 32 communication elements emanating from ground control, against 7 from the airliner. Of these 7, 4 are positive expressions of confidence, 2 are questions and one is a "thank you". Meantime, the coastguard also logged ten interventions over the same period, mainly descriptive and informative (MacPherson 1984). But these were apparently neither acknowledged nor acted on by those on the flight deck trying to save the plane.

This is a not uncommon sequence in both aviation disasters and business failures. The people who manage companies see it as their duty to make reassuring statements, even when, to all others in the immediate vicinity, it is clear that they are "not waving but drowning". There is an almost complete absence of *open-ended requests* for information such as "What's happening?" "How does it seem to you?" "What are we doing wrong?" Yet the answers to these questions would provide useful information for the management to know.

In all hierarchical organizations there is a tendency to rely on what was earlier referred to as "the Biggles factor", and for organizational participants to believe that the great complex system can be managed in extremes by the one man whose name appears at the top of the organizational pyramid (Courtis 1986). But this over-centralization of authority vitiates the chances of effective communication to support difficult or contentious decisions.

As systems become more formalized, they become more rigid, and the vertical and horizontal barriers to communication become more pervasive. Professional and hierarchical segmentation can, in the case of organizations with a strong corporate culture, combine to produce a fortress or siege mentality. This is seen in companies that were traditionally dominant in a particular market that has turned against them. The company acts as if it believes that it owns that market, and its communication system acts systematically to filter out the unwelcome signals indicating that its grip is slipping. The British motor vehicle assembly and motor-cycle industries of the 1960s and 1970s are illustrations of this process.

Sometimes communications problems can be amended only by non-predictable behaviour and the deliberate breaching of accepted protocols of communication, moving from "formal" to "informal" modes. That this is very widespread is indicated by the astonishing success of such books as *In search of excellence* (Peters & Waterman 1982). Not the least illuminating part of the excellence paradigm is its insistence on the importance of "listening", especially to customers, and of what is touchingly called "management by walking about". That is to say, the deliberate search for direct, tactile, sensate information, unmediated by the reams of management paperwork that intervene between the main board and the objective reality of products, sales and cash, is perceived by Peters & Waterman to be something that senior managers typically don't do enough of and need to be reminded about.

Another characteristic syndrome found in the classical bureaucracy type of organization is that of the "over-controlled organization", in which the tightly regulated patterns of rules and regulations create the need for management to rely on communications shortcuts in order to solve minor organizational problems (Weir 1975). Because relatively severe sanctions against minor rule-breaking are in place, these shortcut solutions are not consistently reported on by the decision-making hierarchy. They become part of a pattern of subterranean knowledge about "how things really work here." The organization relapses into a relatively comfortable status quo in which senior management remains systematically ignorant of how lower participants are actually solving their problems in a way that ultimately compromises the security of the system. The Flixborough explosion (Lincolnshire, UK: 1974) indicates the consequences of such a pattern of behaviour, as do studies of computer fraud in apparently highly security-conscious financial organizations such as banks.

Promising opportunities for research

We now need to characterize the sequences of system failure in a way that will be helpful in bringing these two sorts of analyses together, and to develop a vocabulary for discussion of the types of communication failure that will lead to more plausible and realistic accounts of why big planes crash and big businesses fail (Butterworth & Weir 1990).

Large planes crash and large companies fail for many specific proximate reasons. It is important to distinguish the two categories of error defined by Reason (1990) as "active error" and "latent error" (Table 5.2). It is the latter that is evidenced in the normal operation of systems working in degraded mode. So it is necessary to delve back into the history of the failure, interview participants, examine carefully the systems and enquire why it is that these systems seem, in practice, to operate in a way not consistent with the rules and regulations embodied in the manual, and that communication failures are seen to be very prevalent. Typologies of failure need to be established and different histories and sequences of behaviour carefully compared and contrasted. In many specific areas

Table 5.2 Causes of communication error.

Active error
Mistaken intervention
Latent error
Incorrect operating built into system
Amplification error
Vicious cycle of incorrect operating built into system
that subsequently operates in degraded mode
Information about errors and degraded operating
of system is systematically suppressed or distorted

there is now an excellent literature of case material and analysis. One of the most important contributions to our understanding of why large planes crash is Beaty's *The naked pilot* (1991), which shows how limited and organizationally self-seeking are after-the-event explanations in terms of pilot error. A promising field of possibilities for research into the failure of military systems is opened by Dixon's analysis of military incompetence (Dixon 1976).

A first step would be to collect a series of cases, building on the land-mark study of Turner (1978) and developing a typology of the communication processes generated within organizations of direct structural types, possibly using as a basis the types identified by Mintzberg (1979).

From the outset, such a programme of research would need to be inter-disciplinary, involving paradigms from such apparently disparate fields as stress engineering, system dynamics, organizational sociology and criminology (Butterworth & Weir 1990). But without it, our understanding of the complex symptomatology of organizational crisis and catastrophe may always remain anecdotal and the search for an integrating explanatory structure vitiated (Weir 1993).

The concepts of communication and information offer a plausible starting point for the over-arching framework so obviously missing from the Royal Society's report on risk (Royal Society 1992).

CRITERIA FOR THE DESIGN OF
HAZARD MITIGATION INSTITUTIONS

Edmund Penning-Rowsell

Natural hazards and human expectations

Because the public no longer sees natural hazards simply as "Acts of God", they are demanding that society and government, rather than the Almighty, protect them from their effects. Accidents as such no longer exist: someone is to blame.

In this respect, the interface between natural hazards[4] – floods, high winds, earthquakes, and erosion – and other hazards is increasingly becoming blurred (Dynes et al. 1987, Alexander 1993). Populations at risk from natural hazards fail to see these events as naturally occurring phenomena, but instead as something that should have been tackled, preferably at nil cost to them (Green et al. 1991). Private and public authorities are held to be responsible for the impacts of these hazards, and for the hazards themselves, in the sense that "something should have been done" to lessen the vulnerability of modern populations: "Isn't that what we pay our taxes for?".

Thus, people have expectations about risk and impact reduction, and these expectations include that the state will design institutions to bring protection against the unforeseen; insurance bought from the marketplace is not enough. And it is *protection* that is required, not the lessening or alleviation of hazards and risks. This public stance puts pressure on risk-related institutions to "deliver", and to design their structures and systems for this delivery in cost-effective ways (Drabek 1986). The weakness of the conventional engineering approach, which dominates hazard mitigation, is that it purports to meet this naïve aim (or at least the public has this perception), and in so doing loses credibility because absolute risk protection, and designs based on this concept, cannot be delivered. Thus, both the design of hazard alleviation measures and the design of the institutions created in this vein are flawed, because they focus on the "solution" to identified problems rather than the management of vulnerability (see Jones, Ch. 2, this volume). And, of course, this institutional

---

4. The term "natural hazard" poses a problem, since hazards are themselves anthropocentric concepts. However, the term is used here to mean those hazards that have a geophysical basis (floods, earthquakes, storms, droughts, etc.) as opposed to those that have a clearer and dominant human basis (petrochemical explosions, fires, air pollution from automobiles, etc.). For further discussion, see Jones & Hood, Ch. 1, this volume.

design process is relatively uncharted territory, with often only trial and error leading the way.

This discussion will be illustrated with examples from flood-hazard research (Penning-Rowsell et al. 1986, Penning-Rowsell & Fordham 1994), and also the area of designing institutional arrangements for water management generally (OECD 1989), which itself have an important risk dimension in tackling floods, drought, pollution, drinking water quality, and so on. It will also draw on the nature and problems of risk communication, as analyzed in previous writings (Handmer & Penning-Rowsell 1990, Kasperson & Stallen 1990).

The latter emphasis on risk communication is important, because the design of the risk- and impact-reducing agencies should stress and target their internal and external communication systems (Kreps 1989), whereas most water related agencies and utilities are designed to have operational, regulatory or constructional capabilities, and are dominated by engineers. The resultant focus on engineering considerations misses the crucial point that the engineering approach to hazard reduction has limitations, most particularly in creating the impression that problems have been solved rather than lessened, and in not tackling the fact that a proportion of the population therefore continues to be at risk and needs to know of that risk in order to reduce its vulnerability.

Institutional design problems for risk and impact reduction

There are several problems inherent in the design of hazard mitigation agencies, and many are intractable. But analyzing these problems gives some insight into the design task, and gives pointers to likely pitfalls.

The problem of scale

Contradictions here are inherent. Natural hazards tend to be highly localized – although there are exceptions such as droughts and major floods – but the resources and skill mix needed for their management is often only cost-effective when centralized and available in large organizations with responsibilities over a wide geographical area. Institutionally, there is a natural tendency to give prime responsibility to small agencies operating locally, because they will then have intimate knowledge of the hazard problems they face, but it is becoming increasingly recognized that a wider geographical and disciplinary perspective brings a better insight on natural hazards. Moreover, policy inertia means that institutional frameworks, policy instruments, professional compositions and the re-drawing of interprofessional boundaries lag behind the practical needs of hazard mitigation in the modern world. The result is often a messy compromise between local needs and more centralizing structures.

For example, the hazards at the coast – flooding and erosion – are highly localized in extent. It is widely acknowledged that they are often poorly defined and poorly tackled at a local level, yet key responsibilities in Britain remain firmly with small district councils. This is partly inertia, but also partly because local construction-related interest-groups strive to retain influence and are successful in so doing.

Who dominates policy?

Policy-making responds to pressures, from politicians, interest-groups and individuals (Torry 1978, Scanlon 1988). It is also affected by the results of resource allocation decisions: the powerful are those with resources to spend and the support of those who will maintain this status quo.

In the flood-hazard mitigation field, many interest-groups remain powerful, but the groups are changing. There has been a major break in the links between agricultural interests and flood alleviation (land drainage), as the power of agricultural interests has waned as food surpluses have grown. New interest-groups are emerging, concerned with urban land-use control for hazard management, as property developers seek to exploit their land resources by reducing their hazardousness, preferably at public expense (Penning-Rowsell et al. 1986). Sectional interests dominate and public agencies struggle to promote the common good.

The difficulty for the design of hazard mitigation institutions is that those with the problem may not have the power. Flooding problems in British major river catchments should be seen as a product of inadequate land-use planning, such that increased urbanization creates more rapid runoff and land-use control fails to protect floodplains from development (Parker 1995). But the National Rivers Authority[5] remains almost powerless to affect land use, except by exhortation and voluntary policy co-ordination (Penning-Rowsell & Tunstall 1996), because the power to make land-use decisions is jealously guarded by locally elected district and county councils.

Performance targets: what should they be?

A further difficulty with institutional design is "design for what?". The different voices in the natural hazards arena continue to be dominated by the quantifiers and the design optimists. Experience of the natural world and human judgement are eschewed in favour of accumulating data and the quantitative risk assessment process. This means that success in flood defence is defined in terms of water moved, rather than people satisfied.

As outlined above, the prevailing belief is that natural hazard problems can be "solved". This is the dominant engineering view and has the

---

5. Incorporated within The Environmental Agency from 1 April 1996.

limitations that, as with many other hazards, floods cannot in reality be eliminated, only alleviated, yet the public may be given, the impression that they are totally protected when they are not (Parker 1995). Engineering approaches are also capital intensive, resource consuming, slow to implement and potentially environmentally damaging.

Moreover, the dominant engineering and positivistic view is that the solution (and it is usually perceived that there is only one solution) rests on the problem being defined carefully enough (preferably with the latest computer technology) and assumes that budgets are made available for adequate engineering works. The nature of hazards as "people problems", and therefore a product of behavioural responses to factors such as land values, is not appreciated. Modifications to human behaviour are seen as more difficult to achieve than obtaining risk reduction by designing hazards out of our *locus vivendi*, and therefore the construction-related design template for flood-hazard mitigation institutions continues.

As a consequence, retirement colonies at the coast are built in areas liable to flooding and erosion. Property prices in these locations are not depressed by this high hazard potential, because the market anticipates that protection will come. And it usually does, and the responsible institutions are designed accordingly, thereby exacerbating the build-up of assets in vulnerable areas, which will then be used to justify yet more capital investment on engineering schemes in the future. The already-turning concrete mixer is a metaphor that retains abiding value in the spiral of policies of "meeting need" with higher standards of so-called "protection".

Risk communication, warnings and blame

Policy-makers increasingly recognize the difficulty of communicating natural hazardousness to a mobile and inattentive population that is more concerned with day-to-day worries than with infrequent natural events (Handmer & Penning-Rowsell 1990). Attempted solutions promoted by inappropriately designed institutions tend to be a mix of ineffective public education and a reliance on populations reacting "when the time comes". Often they do not, and the result is a complex process of post-event blame allocation.

Indeed, warning and "blame" are becoming increasingly interlinked (Parker & Handmer 1992). The public perception of the failure of the UK Meteorological Office to give adequate warning of the October 1987 wind storm in southern England gave rise to much blaming and the need for acceptance of blame. The same occurred after the disastrous February 1990 Towyn floods in North Wales. Indeed, most flood warning systems fail most of the time (as opposed to the forecasting of flood levels and timing) because the messages are not disseminated to all those who need them.

As a consequence, the messenger is being shot for failing to deliver, and quite right too. But the danger is that this leads to the suppression of self-criticism and the denial of uncertainty. The reason for the failure to deliver in the field of flood forecasting and warning is that all the emphasis is placed on the forecasting of floods and little attention is given to the process of communication with the public. Science and engineering dominate; communication skills are given second place.

The situation is compounded by legal worries. In the case of flooding, some important legal cases (Cardiff and Boroughbridge: see Parker & Handmer 1992) have indicated that those responsible for issuing warnings are more liable than was considered previously to be the case, if the warnings they give are not adequate or perceived as such. In the USA, where recourse to litigation is more common than in Britain, there have also been cases where inadequate storm warnings resulting in damage and deaths have resulted in lengthy lawsuits (Kreps 1992).

The institutional and policy response – at least in Britain – is not just to increase efforts and to enhance the capability to forecast and to warn, but also to limit and delimit liability, occasionally by refusing to issue warnings in case they prove to be inadequate. This is clearly a most counterproductive trend.

Public communication dilemmas: continuous or "when needed"?

In theory at least, the beneficiaries of the efforts of risk-reducing agencies should be involved in their design, but this is difficult in practice because the beneficiaries of increased protection from high-magnitude low-risk events cannot easily be identified in advance of those events occurring.

Also, there is a dilemma concerning the extent to which the public should be involved. Good sense would dictate that potential beneficiaries should be kept informed as to possible hazards that will affect them, so that communication between agency and beneficiary remains good. But there is the danger that repeating messages degrades the value of their content, in the absence of recurring major or minor events, resulting in the "cry wolf" syndrome, whereby warning information is ignored because it is too often repeated.

The alternative is to provide information "when needed", but this suffers from the fact that the receivers of this information will be relatively less prepared and therefore are likely to respond less well. Finding the right balance here is a major objective of hazard mitigation agencies, and yet the basis on which to make this decision is very sparse: institutions are not designed with this crucial calculus in mind.

Standards of service: a way forwards?

One way forwards in the design of flood mitigation institutions is not to define the institutions themselves but to define the service to be provided to meet "customer needs", and then allow this "product led" approach to define the institutions involved. In this way the use of closely defined standards of service would appear to be a solution, but in fact it is a false panacea. Hazard protection is often a public good, and therefore allowing the customers alone to dictate how this service is delivered through charging them for executing this responsibility is difficult, because one cannot realistically allow individuals to opt out of the service provided.

Moreover, the current solution (at least in Britain), of defining a public service and arranging for agencies to compete to provide it, means that one has to define what is acceptable as a service and what is not; this is notoriously difficult in the hazard mitigation field for reasons outlined above. Tying the charges to the service provided is one relatively easy way forwards, at least in theory, and appears to give weight to the needs of the paying customers. The problem is that low-probability events are not perceived by the potential victims as being important enough to pay for, but when they happen there is uproar that they were not protected (the 1995 drought in England and Wales is a classic example of this syndrome).

Despite these difficulties, the moves towards linking public bodies to defined standards of service (or charters, which are the same thing) has given rise to defining standards or levels of service for the "services" that hazard mitigation agencies "provide". The "responsible" authority will be required to define the nature and extent of warning or defence systems, such as the frequency of flooding in a particular area, or the probability of drought or slope failure. This can be such as also to serve as a way of limiting liability – self-protection for the authority concerned – and also to determine the charging systems for the relevant services.

But "service", in terms of hazard mitigation response, is difficult to define for floods that occur infrequently (i.e. once in 100 years or more), and good memory as to what response is appropriate is not encouraged as organizations are constantly reorganized. Given the infrequent nature and the magnitude of these extreme natural hazard events, the reassurance that such standards of service offers is largely illusory, since it is impossible in many cases to determine what standard is being given, let alone what should be given.

Designing better institutions: criteria and choices

Little attention has been focused on the *criteria* required to design an agency for disaster/hazard/risk reduction (be it a fire agency, a flood defence organization, or a pollution-control agency). Surprisingly, research over the past 20 years almost entirely fails to inform on the *choice* of those criteria and their *effectiveness*. Why is this the case?

The research base: less than fully adequate

First, the research. Much of natural hazard research has been located conceptually in the behavioural tradition by examining individuals and their choices, rather than institutions and their dynamics (Parker & Penning-Rowsell 1983, Parker 1992). In this respect, much of the research follows the Gilbert White tradition, which is steeped in American individualism rather than European or British corporatism (Parker & Penning-Rowsell 1983). Solutions to hazard "problems" have correctly been seen as behavioural, but the institutional context has been neglected.

Following on from this, much of the research into the impacts of hazards and disasters and the economics of hazard mitigation (Penning-Rowsell et al. 1992a) has been rooted in a benefit–cost neoclassical economics framework that stresses the importance of individuals in allocating their individual resources and deciding what is optimal for them. Individual utility is the criterion by which resource allocation decisions are made, and only indirectly is the common good approached (and then as the sum of individual goods).

More recent research on non-engineering approaches to flood-hazard mitigation – and particularly land-use control and warning systems (Penning-Rowsell et al. 1992b) – has moved unambiguously into the institutional arena. Understanding the complex webs of inter-institutional relationships is now more clearly seen as the route to the analysis of policy evolution and the enhancement of policy implementation, rather than an emphasis on simple institutional structures and key individuals' interests. Moreover, as research and understanding have increased, the consensus has begun to evaporate on what is suitable and appropriate as public policy in hazard mitigation. Research on hazard/environment trade-offs (Fordham et al. 1989) shows the dilemmas that flow from the fact that natural hazard management requires environmental intervention and modification.

Thus, flood alleviation schemes can adversely affect valued river environments. Coastal protection can harm unique coastal ecosystems. Drought reduction policies often involve supply extension rather than water demand management, and this can require new reservoirs in valued upland areas. So, the traditional medicine of hazard reduction through engineering works is being seen by an increasingly influential series of groups as worse than the disease of the hazard itself.

There are some who advocate trade-offs between hazard reduction and environmental values, using the growing body of environmental economics techniques as a basis for supposedly rational decision-making (see a critique of this in Coker & Richards 1992). Increasingly, there are others who see the problem as essentially human-induced and as a moral question, and for the solution to be changing human use of the environment, rather than changing the environment to suit human needs. This is increasingly revealing different value systems and overturning the pre-existing supposition of an unambiguous common good.

So, research in the hazard mitigation arena has not led to clear lessons about the design of hazard mitigation institutions; indeed, the reverse may be true. There is a need to re-evaluate those institutions and the legal frameworks, administrative structures and policy instruments that they comprise, so that they are made fit for their purpose.

Institutional design criteria

The criteria against which hazard mitigation institutions should be designed (Table 5.3) include their *commitment* to hazard mitigation (given that this may be just one of their functions), their ability to *monitor* and *evaluate* hazard onset situations and their ability to *respond* in hazardous event situations. We also need organizations to *learn* from their experiences and to use this in the redefinition of their policies and practices. The ability to respond will be reflected in the effectiveness of their internal communication systems, the preparedness of their staff, and hence the quality of their pre-hazard training, their ability to communicate with potential hazard victims and advise them on appropriate actions, and their ability to mobilize resources for hazard response activities. In addition, consideration must be given to their ability to *desi n* and *implement* (or assist in implementing) vulnerability-reducing preventive systems. These criteria will be discussed in turn.

Table 5.3 The criteria against which hazard mitigation institutions should be designed.

- Their commitment to hazard mitigation.
- Their ability to monitor and evaluate hazard onset situations.
- Their ability to learn from each hazard experience and use this to determine future policies and operational practices.
- Their ability to respond in hazardous event situations, and thus:
  - The effectiveness of their internal communication systems.
  - Their ability to communicate with potential hazard victims and advise them on appropriate actions.
  - The preparedness of their staff, and hence the quality of their pre-hazard training.
  - Their ability to mobilize resources for hazard response activities.
- Their ability to design and implement (or assist in implementing) vulnerability-reducing preventive systems.

Commitment to hazard mitigation   It may appear self-evident, but the commitment of an agency to hazard mitigation may not be total. This can be because hazard mitigation is just one of many functions undertaken by that agency; for example, one concerned with flood warnings may also have other "water" responsibilities such as water resources, water supply and irrigation. Fire agencies may also be responsible for other "civic" functions such as waste disposal or street cleaning. Police authorities – who usually have a central role in hazard situations – have many other unambiguous definition of responsibilities, there will be uncertainty and "buck passing".

Monitoring and evaluation   Above all, the public requires that hazard mitigation agencies respond in a timely and efficient manner to the hazards they face. This necessitates that the institution has the ability to gather its own information on which to base a forecast of hazard onset, or the ability to obtain such information from others. Thus, agencies concerned with flood hazards have generally employed a network of river gauges, rain gauges, tide and wave gauges at the coast, and such modern devices as weather radar, with which to construct their forecasts of river flood or sea surge levels.

This criterion strongly implies a devolved institutional structure of local activities, designed to identify hazard problems where they occur. Centralization brings a degraded picture of locally hazardous situations, despite modern communication systems. On the other hand, interpretation of hazard onset data is a complex task, often requiring high-powered computing and skilled analysts; it is not a mechanistic operation.

This in turn implies an institutional structure where locally collected data are analyzed and interpreted regionally, or at least not locally, so that regional trends can be observed and point data collated. Thus, the Storm Tide Warning System for the North Sea requires both locally collected data and centralized analysis by a Bracknell-based (southwest of London) forecasting operation. The one cannot operate without the other, and the structure of the agencies involved must reflect this operational need.

The habit of learning   By definition, severe hazards occur infrequently. Each event must be used as a basis for tackling the next event. But this should not mean a ceaseless and restless fluctuation in policies and a random reallocation of duties and resources. What is needed is the proclivity to analyze the performance of the organization in reacting and responding to the events that it tackles, and an inbuilt self-critical learning process whereby mistakes in the past are used constructively to inform future policy shifts and operational arrangements.

The ability to respond to the hazard   The appropriateness of response by an agency to an impending hazard is difficult to specify, as all events are to some extent unique. The need is to define the range of problems likely to be experienced and to tailor the design of the institutions to that range, so that the agencies can respond flexibly to circumstances as they emerge and evolve.

Inevitably, however, good hazard response means that a certain degree of redundancy has to be built into the institutions involved, for it is impossible to plan to mobilize the correct level of resources; some duplication is wise in order to limit the risk of being overwhelmed by an event. This means that communication systems and hazard-fighting equipment may need to be duplicated in case one systems fails, and this applies right throughout the hazard response arrangements.

In effect, the process of hazard response becomes one of linking the allocation of resources for "fighting" the hazard (or alleviating its impact on potentially affected populations) to the nature and extent of the need. This requires good *communication* both within the organization and between the organization and the many other agencies involved in the hazard situation (Fig. 5.1). These communication systems need to be hierarchical, in that the level of decision-making needs to rise as the magnitude of the hazard increases, and they need to be built into the structure of the institutions involved.

Much research stresses that communication systems should be multipath, so that faults can be by-passed (Penning-Rowsell & Fordham 1994). Also, two-way communication links between the agency and the scene of the hazard are important (Handmer & Penning-Rowsell 1990), because in that way feedback loops can be fed with information about the hazard as it evolves, and response tailored flexibly to that changing situation. Also, many hazard situations involve "victims" who have no prior experience of such events, but the agency will have knowledge and experience of similar events elsewhere. Thus, the hazard mitigation agency should serve as the "collective memory" for society and it should be able to advise hazard victims on response strategies. Therefore, communication needs to be established *from* the victims *to* the agency, not just the other way (which is how many hazard agencies see the communication process); indeed the communication process must be designed from first principles as a fully efficient two-way conduit.

A good response to hazardous situations cannot be achieved unless there is good information as to how to respond, and this generally involves the *training* of the institution's staff. Generally, this is well recognized but the implementation is problematic. For example, one of the key problems with tackling low-probability – high-severity events, such as major floods or earthquakes – as opposed to frequent events such as urban domestic fires – is that staff responsibilities for hazard mitigation tend to form just part of their duties, and they also tend to have little experience of actual events (as opposed to simulations). For example, the experience in Britain of dealing with major flood events such as those of 1947 (River Thames) and 1953 (East Coast) is now virtually non-existent and there is much evidence that training is no substitute for real experience (Penning-Rowsell & Winchester 1992).

Standardization of training is important within an organization, as is the standardization of terminology describing the hazard and how to respond (e.g. "red alerts" should not mean different things and therefore

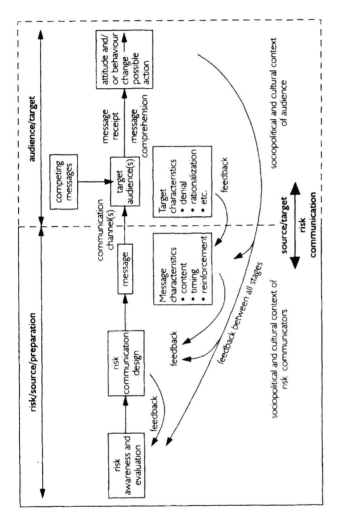

Figure 5.1 The process of risk communication (from Handmer & Penning-Rowsell 1990).

be liable to different interpretations in different parts of an organization). Training needs to be kept up to date, and based on feedback from experience of real events or exercises (Fig. 5.1). Without due attention to these processes, responses will be slow, uneven, and out of date. Training needs to be built into the operation of the organization, rather than regarded as an unnecessary or costly "extra".

Clearly, hazard mitigation also requires resources, and a key function of any hazard mitigation agency is the ability to *mobilize resources* to counter the hazard. This may mean a fire-fighting agency having enough fire-fighting tenders located within a defined radius of any fire, but also the ability to call on other resources at short notice from other agencies nearby or from farther afield as necessary.

This criterion links with that concerned with commitment, and leads to the design of hazard mitigation agencies within a political unit, such as a local government body, whereby political decisions can be made about resource deployment: the mayor is in charge and can mobilize fire-fighting tenders to help with floods, or pollution-control staff to fight fires, and so on. Single-purpose agencies can be weak in this respect, or else they may have to install excessive redundant capacity to deal with the unforeseen. In addition, the line management route above and beyond the hazard mitigation agency must be made clear, whereby the local, regional, and national government can step into a situation when it exceeds the resources of the purpose-designed agency itself; which is in charge of sending in the army, or other emergency forces.

This means that the power and limits to power of the hazard mitigation agency must be clear to those within, "below" and "above" the agency; without this there is a real danger of overlapping responsibilities and degraded performance. Similarly, the nature of responsibility for warning dissemination and response monitoring must be clear, and this will usually mean close definition of the relationship between the specialist hazard mitigation agency and the police or armed forces; this has not always been the case in the past and yet is crucial to appropriate response.

The ability to act preventively   Hazard mitigation centred on the event itself is not enough; a well designed hazard mitigation policy and its institutional base should have preventive policy actions embedded within the agencies involved. However, herein lies a dilemma or at least a weakness. Preventive planning without operational hazard-fighting capability leads to agencies dangerously separated from reality; but excessive concentration on operational hazard-fighting capability leads to the neglect of action to limit the growth of hazard build-up (Penning-Rowsell 1996).

Another dilemma is that the ability to act preventively usually means that the agency has to stray outside its locus of power into areas where it has no statutory or other power base. Thus, in the flood alleviation field, flood agencies need to influence, and thereby constrain, the build-up of flood damage potential by influencing the urbanization of floodplains (Penning-Rowsell & Tunstall 1996). Otherwise, the flood mitigation efforts of the agency will be prejudiced by land-use planning agencies deciding to locate property in areas at risk.

But flood-hazard mitigation agencies usually do not have land-use control powers (except immediately adjacent to rivers or the coast), because land-use decisions are rightly a balance between interests, and flooding is rarely the overriding factor. Therefore, flood mitigation institutions need to operate by seeking to influence those local government bodies that have such powers, and to seek negotiated cooperative solutions (Handmer 1995). Similarly, fire-fighting agencies need to operate by influencing building regulations, road planning, water resources deployment, and a host of other arenas well away from their core business of fighting fires. Yet not to do this means that they are tackling the symptom rather than the cause. This process of negotiation and influence is admittedly difficult and often inefficient, but it is inevitable. Sometimes higher authorities, such as central government, need to coerce hazard mitigation agencies to work more closely with others, and vice versa, but this is rarely the norm and it should not be relied upon in the majority of circumstances.

In terms of institutional design there needs to be a complex web of cross-compliance regulations, such that agencies at least become aware of the activities of others and can seek to influence them. Inter-agency negotiation and cooperation should be promoted as the norm, But if these processes are consistently rebutted or ignored, there needs to be some higher authority that can intervene, which again appears to point to a close association between hazard mitigation agencies and the political domain.

Assessment

It is understood that hazards and risks are complex phenomena. However, we do not understand how to design institutions to tackle these complexities. The state of research and knowledge in this field is antediluvian compared with our knowledge of both the risks or hazards themselves and the way that individuals generally respond.

What experience we have shows that hazard mitigation agencies need to be designed so as to enhance their commitment, facilitate monitoring and evaluation of hazardousness, have the capability to be responsive in a flexible way related to the nature of the hazard and the degree of response required, and also be capable of operating preventively so that risks and hazards do not grow through neglect.

Currently in Britain, no hazard mitigation institutions exist that have been designed using these criteria in order to determine their legal frameworks, administrative structures and policy instruments. They are designed often to tackle yesterday's problems, with ill trained staff, working in quasi-scientific or engineering mode in tackling what is essentially a human problem. The result is increasing hazardousness, poor response and a dissatisfied public. There is much to do to improve this situation.

CHAPTER SIX
Counting the cost

RISK REDUCTION, BUT AT WHAT PRICE?

A fifth area of debate in risk management concerns the cost of risk reduction and the extent to which reduction in risk has to be traded off against other basic goals. The much-invoked doctrine of BATNEEC ("best available technology not entailing excessive costs") explicitly recognizes this trade-off problem, although it does not tell us how to choose among competing interpretations of what is "best available" and "excessive".

The conventional "no-free-lunch" trade-off model of economics offers a clear starting-point for risk management, since it focuses attention very sharply on discounted costs and benefits. Those who adopt a trade-off position argue that increases in safety must normally come at the expense of other valued objectives such as wealth creation, international compet-itiveness in productivity or economic dynamism, and environmental deg-radation. Breyer (1993: 11–12) argues that, in the absence of attention to costs, attempts to regulate small but significant risks to human health tend to be characterized by "tunnel vision" (which he see as arising "when an agency so organises or subdivides its tasks that each employee's indi-vidual conscientious performance effectively carries single-minded pur-suit of a single goal too far, to the point where it brings about more harm than good" (ibid.: 12), for example in disproportionate expenditure and effort to remove the "last 10 per cent" of hazards. The debate over the costs of risk reduction has included discussion of the economic con-sequences of extended product liability in the USA, although there is no clear overall evaluation of the economic effects of those legal trends (Reu-ter 1988). At the limit, the cost of risk reduction can be framed not simply as the trading-off of risk reduction against other competing values, but also as the trading-off of some kinds of risks against others (e.g. botulism risks versus cancer risks in relation to nitrites in food, or risks of dumping sewage sludge at sea versus cancer risks from incineration on shore).

Much of public policy on risk management can be represented as an implicit trade-off between safety and economic surplus. By the use of analytical techniques designed to make those trade-offs explicit, an effort can be made to assess the consistency with which the trade-off is made in different areas of public policy, and why it varies from one case to another: for example, why the trade-off appears to be different for aviation and railways as against road transport.

Against this orthodox view is an alternative position, which holds that high safety standards can be achieved in conjunction with other goals, and also that good risk management is one of the signs of good management in general. The claim here is that badly managed organizations are likely to have poor safety records, along with poor performance in other dimensions. From this standpoint, increased attention to safety can be seen to pay for itself, since prevention through good design, good training and good practice will prove to be cheaper over time than the costs incurred by disruption, damage claims and insurance premiums, poor public image and a loss of goodwill from employees (cf. Kloman 1990). Such arguments have figured prominently in the justification for the wide-ranging corporate safety programmes and have even been enshrined in the OECD chemical industry guidelines for the handling of hazardous materials.

In the case of environmental pollution, Tait & Levidow (1992) argue that regulation that requires developments in technology does not necessarily come at the expense of profits and that "where such an approach has been adopted, as for example in the German car industry, there is no evidence that it has disadvantaged the companies concerned", and the RCEP Report on best practicable environmental options (1988: para. 1.12) points to "sound business sense" arguments for increased efficiency and waste reduction through better environmental management. Similarly, in relation to crowd safety, some have argued that expensive safety-related investments in physical facilities can achieve compensatory returns, as an increasingly safety-conscious public comes to select venues according to their levels of safety provision, and perhaps is prepared to pay higher admission prices in return for improved facilities.

The debate over the extent to which risk reduction needs to be traded against other values, and how such a trade-off should be made, was a notable feature of discussion on UK transport safety in the 1980s, and the issue is also very important in the safety debate for hazardous process industries. The debate turns on the precise shape of industrial production functions and of the risk-related characteristics of broader technical, ecological and macroeconomic functions. Up to now, the view that risk reduction can come without sacrifice of other major goals has been more in evidence in debates over risk management by corporations than in public management more generally. It remains to be seen whether further research can elucidate the shape of the social and corporate production functions that are in dispute in these debates.

These contentious issues are explored further in the following two sections. First Tom Horlick-Jones discusses the possibilities of reducing accident rates through the adoption of total quality management. He argues that a safety culture approach to risk management offers a promising means to build safety into corporate management and operations, although much research is required concerning exactly how such an approach is best implemented within established corporate structures. Sir Christopher Foster then assesses the extent to which a safety economics approach can feasibly achieve consistency across the diverse field of risk management and, like Breyer (1993), sounds a warning that excessive investment in safety can prove counterproductive.

IS SAFETY A BY-PRODUCT OF QUALITY MANAGEMENT?

Tom Horlick-Jones

Introduction

Risk management, by definition, is a matter of trading-off harms and inconveniences against costs. All human activities entail risks – to people, to the environment, to bank balances, and sometimes to all three. This section explores the extent to which one possible output of any enterprise – safety – may be achieved without unacceptable losses to other outputs, such as competitiveness or profits.

Hood et al. (1992) have identified a tension in risk management thinking between what they call "trade-offist" and "complementarist" positions. According to this categorization, the former position recognizes the need for safety to be traded-off explicitly against other goals, whereas the latter holds that safety and other objectives may be achieved jointly, this being made possible by good management practices.

To some extent, this assumed "debate" is a rather contrived way of looking at problems concerning the cost of risk management. In practice, although expenditure on safety-related activities is an important consideration, the safety record of a given enterprise cannot be controlled in any simple way by adjusting such expenditure. Safety involves a multidimensional interaction of human, managerial and technical factors with a host of influences from the operating environment. Therefore, although a certain level of expenditure on safety is a necessary condition of safe operations, it is not a sufficient one.

Public regulatory policy on safety may be represented in formal terms as such a trade-off between safety and costs, utilizing techniques that purport to provide a rational, quantified determination of a publicly acceptable balance of harms. However, the implementation of such policy is a matter for corporate management, and corporate behaviour is far from determined by the intentions and actions of regulatory bodies.

So, perhaps this issue can more accurately and fundamentally be represented as a management problem. Regulatory agencies may set safety standards and seek to influence corporate behaviour using a variety of means; however, the achievement of safety is embedded within the complexities of the management process itself. Here it competes with, and sometimes complements, the achievement of other objectives.

This section examines the advantages of adopting quality approaches to management, in which a zero-accident rate is one objective, rather than basing risk management on seeking to limit risk exposure to some afford-able and tolerable level.

The dialectics of safety management

As discussed in Chapter 3, all disasters are sociotechnical events in nature (Turner 1978, 1994, Horlick-Jones et al. 1993, Toft & Reynolds 1994), with human, organizational and managerial features playing sig-nificant roles, often in complex relationships with the technical charac-teristics of the associated failure[1]. Similarly, all manifestations of risk emerge as outputs from such systems associated with human activities.

Safety-related expenditure on the robustness and reliability of the technical infrastructure of any given enterprise is essential, in the form of maintenance, for example. However, technical factors tend to play a relatively minor role in disasters in comparison with managerial and human factors, which are implicated in some 70–80 per cent of failures (Turner 1994).

It is important to stress that these factors cannot be considered in iso-lation. Technical, human, managerial and cultural dimensions interact in a contingent open-ended process that precludes deterministic analysis (Bijker & Law 1992). The operational characteristics of the technology in question will pose varying degrees of challenge to the establishment of safe operating procedures (Perrow 1984, Thomas 1988, Collingridge 1992). However, the most effective means by which such safety manage-ment can be implemented, combining organizational design, corporate cultures and managerial practices, is a matter of continuing debate (Sagan 1993).

The safety characteristics of a given sociotechnical system may be affected, and potentially compromised, by the influence of that system's operating environment, including socio-economic, regulatory and other factors (Horlick-Jones 1990, Horlick-Jones et al. 1993). Indeed, the turbu-lence associated with changes in the British political economy in the 1980s led Dr (now Sir) John Cullen, then Chairman of the Health and Safety Commission to state (quoted in the *Guardian* 1989):

The enterprise culture, the opening-up of markets, and the need to survive competition place businesses under unprecedented pres-sure . . . the scale and pace of technological change means that increasing numbers of people – the public as well as employees – are potentially at risk.

1. It should be noted that this section is not concerned with "natural" hazards.

Such considerations form part of the basis of recent controversies in Britain concerning plans to move the railway and nuclear power industries from public into private ownership. Opponents of the "privatization" argue that commercial pressures may lead to "safety being sacrificed for profits". In addition to the very real safety concerns, such accusations are powerful political weapons in the hands of those opposed to privatization for a range of other reasons. Indeed, the continuing controversies may have an adverse affect on the success of the selling-off process itself, a fact fully appreciated by critics.

Even in an increasingly market-led safety regime, public policy and corresponding regulatory activity introduce important influences on the dynamics of these sociotechnical systems. Wynne (1987: 4) describes these processes as:

> . . . a continuing multi-organizational process of interaction, negotiation, commitment and adaptation.

Again, it is important to stress the open-ended nature of such processes. Having set benchmark standards, regulators use a variety of tools, from persuasion to coercion, to steer the behaviour of management, where they compete with commercial and other pressures on corporate policy and practice.

Public safety policy itself, of course, emerges as a political compromise between calls for strong regulation and arguments that over-regulation threatens wealth creation and suppresses the market's self-regulatory "hidden hand". In the latter lobby, Wildavsky (1991), for example, has pointed to the negative and stultifying features of regulation, making the case for dynamic risk-taking as a more effective road to safety.

Given the interactive nature of the processes that influence safety, and the uncertainty and controversy that surround these issues, what approach emerges as the most satisfactory means of achieving safety? First we will examine the "orthodox" trade-off between safety and costs.

What is an acceptable level of safety?

Pursuing an approach to safety management that seeks to trade-off safety against costs begs some fundamental questions. In particular, how is an acceptable level of risk exposure to be determined? How can a balance of harms be established between risk exposure and the cost of safety measures?

Arguably, the balance between the economic benefits of North Sea oil and the lamentable safety record of the industry was, for many years, unacceptable (Carson 1982). A decade or so later, following the 1988 Piper Alpha disaster, a new safety regime was being implemented with, according to some (*The Times* 1990), severe consequences for production and, presumably, profits. Similarly, it was recently claimed that a pre-occupation with safety has resulted in nuclear power being "over-engineered" (Hay 1991) and that over-conservative safety arrangements are actually hampering performance. Given the contentious nature of these debates, can a rational approach to balancing harms be found?

In Britain, regulatory agencies have adopted risk-control frameworks that, in various ways, involve a trade-off between what is acceptable and how much that level of control costs. As the Royal Commission on Environmental Pollution (1988) has observed, the concept of "best practicable means" (BPM), interpreted as involving a balance of these factors, has its roots in the 1874 Alkali Act, and has formed the basis of air pollution control since that time.

The related terms "reasonable practicability" has been built into the HSE's most recent statement of its risk management approach (HSE 1992a) through the principle of "as low as reasonably practicable" or ALARP. A similar principle, "as low as reasonably achievable", or ALARA, has been used in the nuclear industry for some years as a basis for limiting radiation exposure. Although, as O'Riordan (1987: 208) has noted, "the (nuclear) industry prefers to think there is no difference between the two", an important difference does in fact exist. The adoption of ALARP signals a more explicit recognition that safety may come at the expense of other goals, transcending a position where a tacit decision is made on the basis of what measures are technically feasible and how much such measures cost.

What exactly constitutes "reasonable practicability"? The term has been explored in the courts (a 1949 case concerning mine safety, as cited in Royal Society Study Group 1983: 184) and found to be: ". . . a narrower term than "'physically possible'"

The judge in question went on to state (ibid.) that:

> . . . a computation must be made in which the quantum of risk is placed in one scale and the sacrifice involved in the measures necessary for averting the risk (whether in money, time or trouble) is placed in the other, and that, if it be shown that there is a gross disproportion between them – the risk being insignificant in relation to the sacrifice – the defendants discharge the onus upon them.

In other words there is a presumption of safety, and the burden of proof lies with the objector to demonstrate unacceptable risk. Clearly, this raises difficult issues concerning the social and power relationships between those who are making the sacrifice and those who are being exposed to the risks (see, for example, Ravetz 1990), and questions concerning the plurality of risk perception and acceptability (Pidgeon et al. 1992).

A significant shift in official recognition of the plurality of risk acceptability came with the publication of Sir Frank Layfield's report on the public inquiry into the proposed construction of the Sizewell B nuclear power station (Layfield 1987). This introduced the concept of "tolerability" of risk, an assessment including a risk evaluation that takes account of public opinion, and an explicit assessment of the benefits of the corresponding activity. However, as Kemp (1991) has observed, arrangements allowing a move away from expert judgements and limited public involvement in decision-making have been slow to be introduced, despite Layfield's recommendations.

Supposedly "rational" approaches to establishing acceptable balances of harms do exist. These utilize formal techniques to quantify risk and to evaluate, in monetary terms, the value of entities such as human life and the environment. The merits or otherwise of quantified risk assessment are discussed at length in Chapter 4 of this book. It suffices to observe that various sources of uncertainty, above all arising from human factors and sources in the operating environment, pose severe difficulties for these techniques.

Methods that seek to elicit the value of detriments, whether to the health and safety of workers, or the public (on such "value of life" techniques see Marin 1992a), or damage to the physical environment (e.g. Pearce et al. 1983), have also been developed. Broadly speaking, they either inquire directly into an individual's "willingness to pay" to reduce risk or, indirectly, by observation of actual expenditure, or "revealed preferences", on protection while engaged in other risk-bearing activities.

Arguably such methods remove risk judgements from their specific social contexts. This point has been made, perhaps most strongly, by Wynne (e.g. 1992), who recognizes the role of trust in, and dependence upon, institutions as central to determining the outcome of these surveys. The essentially political nature of risk means that tolerability emerges as an unstable result from both overt and underlying processes of negotiation.

In practice, a range of actual cases demonstrate that the occurrence of disasters can call into question the adequacy of these quantitative techniques. Hence, as O'Riordan & Wynne (1987) have noted, the investment considered necessary to save a life can spectacularly exceed insurance industry standards and compensation arising from court cases. For example, the estimated £20 million per life saved in preventive safety measures following the 1968 Ronan Point building collapse disaster in London (Royal Society 1983).

Similarly, in the case of the Ford Pinto car (see Kleindorfer et al. 1993), the political nature of risk overwhelmed supposedly rational approaches to its management. Here, a leaked memorandum setting out details of a decision, based on cost–benefit analysis, not to make an $11 modification that would improve fire safety in crashes, resulted in damaging publicity and many legal damage suits. Subsequent models featured the modification irrespective of the "economic rationality" of the decision.

Such processes can reverberate around the corporate environment, as demonstrated by the Bowman & Kunreuther (1988) study of the impact of the Indian Bhopal disaster on the behaviour of a chemical company. Here, concern about future Bhopal-like events resulted in the company shifting its risk management procedures from determining "acceptable" levels of risk to ensuring the prevention of worse-case scenarios.

There is even evidence that, despite the formal utilization of quantified cost–benefit approaches, the practical implementation of safety may be achieved by other, informal means. Hence, as recognized in their study of nuclear risk management, O'Riordan et al. (1987: 368) note that:

> It is at the margin of risk management, i.e. where safety investments, technologies and management practices are judged to be satisfactory, that engineering and regulatory judgements approach the weighing of gains and losses rather differently . . . design engineers do not believe that risks can be accumulated into single numbers or that any given safety investment reduces collective risk. They look at design parameters and their implications for operator error and accident sequences. They rarely concern themselves with any formal economic evaluation.

Perhaps most important among the shortcomings of the explicit trade-off approach to risk management are the messages it sends to an enterprise's workforce. This issue has been examined by Needham (1992), who argues that employees "read" management attitudes towards safety, and provide what they perceive management to want. He notes cases where mixed signals have been given by management concerning the balance between safety and production, despite individual managers' good intentions, because of clear "production still takes precedence" messages.

Clearly, the "orthodoxy" of trading-off safety against costs presents serious intersecting difficulties: the technical problems of quantified risk assessment, valuation methods that yield results which evaporate in the light of disaster, and managerial processes that generate uncertainty and ambiguity. In the next section attention will be focused on approaches to risk management that place safety at the heart of the management process.

The quality management approach

One would suspect that industry hardly needs to be reminded about the costs of failure. Yet this message has formed the basis of a recent official safety campaign in the UK, which notes that the costs of accidents can lead to major losses, perhaps up to 37 per cent of profits (HSE 1990a, 1992b, 1993). Some possible effects of safety failures are, of course, difficult to quantify: reduction of share value, damage to public image, loss of market position and so on. As Kloman (1990) has observed: ". . . the most important resource of any organisation is its public reputation, and the most serious risk in the marketplace is loss of credibility".

The difficulty in quantifying or, indeed, controlling these possible losses has led to significant interest in approaches to risk management that seek to prevent or minimize losses. Environmental risks, in particular, have generated much discussion in the business community with concerns about possible post-disaster losses in share value (e.g. *Independent* 1992a) and a recognition that "green" policies may improve a company's public image (CBI 1986, Deansley & Papanicolaou 1992). Indeed, Cairncross (1991) has argued that in future the most successful companies will pursue vigorous environmental policies, in response to incentives set by governments under pressure from concerned citizens.

To what extent, though, do such policies compromise corporate performance? Perhaps surprisingly, research into the structural response of enterprises faced with the management of specific chemical risks (Zimmerman 1985), or with a more general range of environmental threats (Groenewold & Vergragt 1991), reveals a positive tendency towards the encouragement of innovative design of products and processes. Tait & Levidow (1992) argue that where such changes have taken place, for example in the German car industry, no evidence exists of the changes being to the company's disadvantage.

A recent survey of the corporate practices of some 200 firms carried out by the Confederation of British Industry (CBI 1990: 4) revealed that high standards of health and safety contributed to sound management by "contributing to excellence in quality and service (and) by establishing a reputation as a business in control of the risks it creates". In practical terms, the CBI went on to argue for the adoption of a "safety culture" approach to risk management that stresses active workforce participation, team spirit, and training and performance monitoring. Such ideas are familiar components of total quality methods for corporate management (Peters & Waterman 1982, DTI 1991). They also correspond with the features of organizational subcultures recognized to promote safe operations, namely well trained employees, rewards for identifying and reporting problems, and responsiveness to such findings (Turner et al. 1989, Turner 1991, 1994).

Total quality management (TQM) embodies management techniques that seek to generate a highly motivated workforce, committed to producing high-quality products and preventing failure. There is considerable evidence that these approaches, focusing on participative corporate cultures and "corporate learning", do indeed produce quality outputs and have high performance reports (e.g. Denison 1984, Forward et al. 1991). Arguing from a perspective encompassing a very wide range of industrial risks, Needham (1992: 325) points to the advantages of a TQM approach that: ". . . involves the workforce and encourages a flexible and questioning attitude on the part of management and employees alike". Increasingly, the ideas of TQM are being applied not only to improve general corporate performance but also, through the related concept of "safety culture", as a means to manage risks, especially ones of high consequence. Industries involved in managing risk associated with nuclear reactors (ACSNI 1993), software (Schulmeyer 1990), process hazards (Cacciabua et al. 1994) and oil transportation (*Lloyds List* 1994, *Seatrade Review* 1994) are benefiting from the advantages of management processes that seek to achieve accident-free operations, rather than accepting a "tolerable" level of failures.

The example of Shell's marine operations cited above (ibid.) is particularly impressive. Here, the establishment of a safety management system some 15 years ago has corresponded with a thirtyfold decrease in injuries involving lost time on the company's tanker fleet.

Despite some clear advantages, the quality management approach to safety suffers from shortcomings that require particular attention in future research and development. First, as noted by Fortune & Peters (1995), although quality management may be very effective in improving reliability and performance, it may be less effective at identifying high-level high-consequence problems such as the complex latent failure pathways that can lead to disaster.

Secondly, it has been recognized that the implementation of TQM can be carried out in such a way that the bureaucracy of auditing, or perhaps the process of attaining some formal quality accreditation, can become more important than generating quality practices (*Financial Times* 1994, Power 1994). This is a very real danger and it is indicative of how difficult a task it is to change corporate subcultures. More specifically, there may exist considerable difficulty in generating changes in corporate cultures "from above". Research has recognized the subtlety and resistance to manipulation of these entities, and that safety culture is not something that can be "bolted on" to an organization (Turner et al. 1989).

Finally, a certain amount of investment is required to implement safety management systems and the training and auditing procedures associated with TQM and safety culture approaches to risk management. Difficulties may therefore be created, especially for small and medium enterprises that

may possess insufficient in-house expertise and resources to introduce
TQM effectively (Cacciabua et al. 1994, *Seatrade Review* 1994).

Travelling hopefully – rail transport safety in the UK

The case of rail safety serves to illustrate some the ideas discussed in this
section. This is perhaps the most contentious area in debates within the
UK concerning the relationship between safety and costs. The continuing
fragmentation and privatization of the rail network provides the natural
focus of attention for several distinct risk management themes that interact in untidy and sometimes contradictory ways.

Let us first consider an historical perspective. In 1977, during the
delivery of the Fifteenth Sir Seymour Biscoe Tritton Lecture, Captain
I. McNaughton, the UK Chief Inspecting Officer of Railways, argued that
less could legitimately be spent on rail safety. In his view (McNaughton
1977: 9) the level of rail safety at that time was "an acceptable one, and
sufficiently above the minimum standard acceptable to public opinion to
allow the occasional serious accident to occur without undue reaction".
This led him to conclude that (ibid.: 1) "some relaxation of the current
safety requirements, with consequent cost savings, might be made without a significant lowering of safety standards". A decade later, Sir
Anthony Hidden's public inquiry (1989: 133) into the Clapham rail disaster warned that "more could and should be done to ensure that safety
is not compromised by permitting commercial considerations to delay
investment in safety-related projects". Notwithstanding the important
political changes that had occurred during the intervening years, these
observations illustrate the fact that attitudes towards safety standards, the
cost of safety and the regulatory orthodoxy of the day, can shift dramatically in a short space of time.

The Clapham rail crash in 1988, in which 35 people were killed when
a signal failure resulted in a commuter express train ramming a stationary
one, implicated inadequate maintenance as a contributory factor in the
disaster (ibid.). The crash may well have signalled the vulnerability of a
railway system that had experienced reductions in investment over a decade or more, had become increasingly busy and complex, and yet still
possessed rather old-fashioned approaches to safety management (see
discussion in Horlick-Jones 1990). Indeed, British Rail (BR) admitted that
before the Clapham crash its approach to safety was based exclusively on
the reliability of equipment (*Independent* 1990).

What happened at Clapham and subsequently at Purley (see Ch. 3)
demonstrated conclusively the inadequacy of BR's approach to risk management. A safety management programme has now been introduced

(ibid.), yet it is important to recognize the difficulties of creating culture changes in an industry experiencing the fragmentation and turbulence of increasing exposure to market forces. In this sense, the possible threats associated with "privatization" already exist to a substantial degree.

Also, in the railway system's operating environment, the impact of regulation raises some interesting points. The 1988 Fennell inquiry into the King's Cross underground railway disaster has called for the Railway Inspectorate to be "brought up to establishment" and that it should be "vigorous in the discharge of its duties" (quoted in Horlick-Jones 1990: 24). In 1991, significantly in an era of deregulation, the Railway Inspectorate was moved from the Department of Transport to the HSE, drawing a clear institutional divide between the economic interests of the railways and its safety. This change was one of two such transfers, the other being the post-Piper Alpha disaster Offshore Safety Inspectorate (HSE 1992d).

Prompted by the threat of terrorist attacks, and of the disruption from associated false alarms, British Rail and the 1992 HSE's Appleton Inquiry (HSE 1992c) into safety on the London Underground railway system used quantified risk assessment (QRA) techniques to rank the risks posed by fire, and by other threats including train collisions and the need for underground evacuations. The latter study concluded that disproportionate resources were being spent on fire safety measures, a response to the King's Cross disaster in 1987 (*Independent* 1992b), in comparison with other safety measures. Application of QRA, the report argues (HSE 1992c: 16), avoids "the possibility of wasting considerable amounts of public money".

The clear emergence of a such a quantified cost–benefit approach to rail safety is perhaps unsurprising in an industry so aware of its diminishing subsidy from central government and its need to be seen as competitive in economic terms. This has clearly been the case, as demonstrated in the increasingly strained debate concerning the proposed introduction of automatic train protection (ATP) (see discussion in Hamer 1995). This safety measure was recommended by the Clapham disaster inquiry (although it would not have prevented that particular disaster) and would have prevented several fatal crashes, including the ones at Purley (1989) and at Cowden in Kent (1994).

Although no final decision has yet been made, it seems likely that the considerable cost of introducing ATP across the entire rail network will be seen as unjustifiable in comparison with the extent of loss of life and damage that it would prevent (ibid.). The associated logic of economic rationality ran through the proceedings of a recent industry conference, suggesting very little in the way of alternative approaches to managing the risks (see Evans & Maidment 1995).

Safety in the UK rail system emerges as an output from complex changes in technical infrastructure, operating procedures and management styles, impacted upon by shifts in funding and regulatory regime, and by the powerful influence of tragic and newsworthy disasters. Risk management is carried out by means of a pragmatic combination of safety management systems and cost–benefit trade-off calculations. Whether this approach will prove effective in coping with the challenges of the next few years remains to be seen.

Conclusions

Cost–benefit techniques for managing risk, which seek to establish a value for loss of life or other harms, treat risk as if could be measured and calibrated like some physical quantity such as temperature. Such approaches fail to capture the social, cultural and political nature of the risk construct, demonstrated perhaps most effectively by the nature of a disaster's impact (Horlick-Jones 1995), and by its subsequent consequences on corporate behaviour and operating environment, and on future performance.

A quality management, or safety culture, approach to risk management offers a promising means to build safety into corporate planning and operations. In particular, it generates corporate practices and cultures predicated on the prevention of failures, so avoiding the negative and difficult-to-quantify consequences of disasters. This approach avoids the need to specify a tolerable level of harm and the technical and management problems of seeking to implement risk management on this basis. This philosophy and practice is compatible with "preventive" risk management associated with the "precautionary principle" which has increasing influence in environmental regulation (Wynne 1992b).

Despite the promise of this approach, much more research is needed into addressing the shortcomings of, and developing, the existing means of implementing TQM and safety cultures, in particular in complex corporate environments such as those presented by the UK railway system.

RISK MANAGEMENT:
AN ECONOMIST'S APPROACH

Sir Christopher Foster

Introduction

At various times in my career and in various capacities and circumstances, I have come across techniques of rational risk assessment, but I am not an expert on them. Although I have frequently encountered issues that require the use of such techniques, I regret that most people know more about them than I do. So, my function in this section is to be provocative. The manner of my provocation is to argue that, although it is admirable to rescue so large a class of decisions from politics and prejudices, one can go too far with it to the public detriment, unless economic considerations are taken into account.

Safety is but one factor among many important factors, such as impact on the environment, which are increasingly recognized as requiring to be evaluated in decision-making. One way of looking at the issues I wish to discuss in this section, is that they raise the question as to what weight should be given to such factors in decision-making. As I see it, that is what the economics of risk assessment is about.

For example, there is no question that a situation that causes, or may cause, an accident is a proper object for concern; but equally it ought to make one pause for thought when, as I recall, the leading counsel for the CEGB at the Sizewell Inquiry stated that, in their expert opinion, if Sizewell B had been built at the time of Queen Boadicea (Boudicca), there would have been only one extra fatality in its vicinity since. In my judgement, such a statement should make one wonder if the emphasis on safety expenditure may not have been taken too far.

It seems quite common to believe that the two main obstacles to the elimination of accidents in any given set of circumstances are the limitations of available technology and questions of personal freedom. As far as the first is concerned, it is only a matter of investment in the relevant research and then the passing of time that decides how quickly accidents of a particular kind are eliminated, whereas the second obstacle leads one into a forest of ethical questions that, in the end, are normally decided by the political process.

Instead, my argument is that, often prior to this, it frequently makes sense to ask if it is appropriate to commit more resources to reducing a particular class of accidents. In practice it is scarcity of funds available to adopt every existing and potential technical improvement that, whether one likes it or not, makes economic factors the actual limiting obstacles to progress.

Safety economics

The purpose of safety economics is primarily to address the question whether it is feasible, rather than actually possible (a more demanding test) to achieve consistency within such areas of concern as road safety, or between that and rail safety, or between such areas as nuclear and coal electricity generation, so as to use resources cost-effectively.

One may contrast an economic approach to achieving consistency with a legalistic one. For example, there may be an attempt to formulate rules. In practice the difficulty here seems to be that inconsistency results, because rules are normally evolved by different regulators in different fields, or even in the same field. Another legalistic process at work results from the many inquiries – for example, the Hidden Inquiry (1989) – set up to investigate particular safety incidents. Normally they reach many conclusions, which are often set out as detailed recommendations for action. One cannot expect this to lead to any great consistency in rule-making. What is sensible in one set of circumstances may prove to be extreme, insufficient or simply inappropriate in others. Yet there is a tendency to use these recommendations widely. Similarly, safety legislation often appears to produce requirements that seem inconsistent between fields of application, again because laws are produced at different times following different initiatives.

A quite different approach to consistency is to use risk assessment. That risk assessment should be done as part of the decision process seems to me incontrovertible; but to rely on it alone risks making it do more than it can. For example, a definition that proposed that one went on taking measures to reduce risk until risks were equal might have an *a priori* plausibility, but it would be seriously flawed, even absurd, because of what would be omitted. However, a developed notion of such an approach is found in the maxims often enshrined in UK legislation: as low a risk as is achievable (ALARA) or practical (ALARP) or "best endeavours at not increased cost" (BEATNIC). These maxims put some brake on the equalization of risks as an objective but leave unclear what these factors are or what weight is to be given to them.

My contention is that all the methods listed above, either in theory or in practice or both, give insufficient, erratic or no weight to cost considerations and are therefore incomplete. As a result, they do not give weight to the important policy consideration of deciding how far one should go in reducing accidents. Adoption of ALARA, or another of that family of criteria, has the result that how far safety is to be given priority is often decided on the basis of what a particular official thinks it reasonable to achieve at a point in time. Moreover, as time passes and what is achievable increases, it reinforces a tendency to spend to an extent that can arguably become disproportionate to cost.

These approaches, therefore, do not permit an answer to any of the following questions. When do diminishing returns set in? When should one stop spending more on safety? Should one spend more on road safety or nuclear safety? Even to raise or question the conclusion to the last question needs a different approach. All such questions require economic analysis, as also does the frequent need to relate safety to other benefits and costs. One example would be where a safety investment has environmental benefits as well. Another would be where an investment just fails the test on normal commercial criteria but will produce safety benefits as well. What difference should those additional benefits make to the decision to go ahead? To be rational requires making such different benefits and costs commensurate.

This is not to deny that risk assessment is also relevant to rational decision-making on safety. Objective risk assessment would seem easier in some areas where there is substantial experience to go on. One knows that nuclear risk assessment is highly dependent on experience of the early nuclear bombs and trials. Elsewhere one often wishes that better records had been kept, but in many industries there is some recorded history of incidents that may be used to supplement, or even supplant, subjective assessments. Where subjective assessments alone are possible, again improving their quality is possible through training and through the adoption of various techniques. Thus, rational decision-making is as incomplete without risk assessment as it would be if it relied on it alone.

As well as risks, one also needs assessment of the relevant uncertainties, for example regarding the effect the various solutions will have on risks. In practice, even where objective assessment of current risks is possible, one is nearly always forced back on expert opinion, that is on subjective assessment, when it comes to estimating the effects of changes in policy, or in investment.

A further area where systematic assessment is possible is over the decision any organization takes to determine its degree of risk aversion. One does not expect any safety organization to declare itself risk prone, but its degree of risk aversion may well vary.

It is at this point that one should again introduce the thought that deciding on a degree or risk aversion cannot sensibly be done except in relation to the availability of resources. Indeed, risk aversion seems to be more often than not shorthand for deciding the relevance of economic factors to safety policy and investment; since if there were not a price-tag to reducing risks, the question of risk-aversion would not arise. Or to put it another way round, the more abundant the available resources, the easier it is to afford to be risk averse.

As a matter of good practice, one should never consider a measure that is predicted to reduce risk without at least a ball-park estimate of its cost. In my experience one often finds discontinuities: measures that are cost-effective in a particular area of concern, then a break before further measures, which are substantially less cost-effective, either because their cost rises sharply or their risk reduction diminishes.

Thus, even if one were not prepared to take the step I am about to recommend, I believe it would be worthwhile to undertake studies of cost-effectiveness. To examine, for example, the cost of different measures to reduce fatalities within and across different fields would, I am sure, show up many inconsistencies in practice. It would indicate where one could save costs by a switch in resources without reducing outputs, that is increasing fatalities. However, such cost-effectiveness studies will not allow one to ask the question: when or where should one stop spending money on safety, because the marginal benefit from doing so falls below the marginal cost. There is no difficulty in doing this for damage-only accidents. No one could surely dispute that an improvement to a car that at a cost of £1000 reduces the near certainty of incurring damage over the lifetime of the car by more than £1000 net present value, is worth doing when all the relevant interests are considered.

Similar considerations arise when the measure is predicted to reduce the frequency of major or minor injuries to employed persons. The loss of production to the employer, or of income to the employed or to whomever else incurs a cost as a result, provides a measure of the gain, albeit a minimum one, from the reduced frequency of accidents. To incorporate that in an economic evaluation of a measure predicted to reduce accidents will give some indication of the relationship between its marginal cost and marginal benefits, although it makes no allowance for the pain and suffering experienced by the injured or by others close to them. Again, for the scaling purposes for which we are using economic evaluation, it is often common and sensible to add an arbitrary mark-up to the measured cost or loss of income to reflect these intangible costs.

The value of life

However, most important questions of safety involve fatalities. If one is to be radical here, one needs to think what may seem unthinkable to many, because it is a logical necessity that one sets a value on a human life. To show this, one only needs to reflect what would be implied if the value of a human life were infinite, not in religious terms, but mathematically. It would imply an infinite return from any investment that saved a single human life. It would follow from this that all society's investible resources should be devoted to nothing else, since nothing else delivers an infinite return.

If it is accepted that this is unhelpful, if not absurd, and accepts that a final value must be given, then one is on the slippery slope. A useful approach is to invert the problem. For example, at the Sizewell Inquiry it was often possible to work out the implied value of a life. One knew the cost of a measure and the predicted reduction in fatalities through risk analysis. Then, assigning a reasonable rate of interest for discounting purposes, one could work at the value of a life needed to make positive the returns on the investment. At Sizewell such values varied at least from £2.5 million up to as high as £25 million for electromagnetic filtration. By contrast, the UK Department of Transport has customarily used figures of between £600 000 and £750 000.

In my judgement such wide spreads of values are indefensible. Human lives are not variously valuable depending on the circumstances of their death. It is sometimes argued that one ought to put a greater value on a death – that is, be ready to spend more money in preventing it – if there is a greater likelihood that there will be a large number of deaths simultaneously, or if the deaths are particularly painful or otherwise unpleasant. Another reason given for different values is that values should be higher where government is seen as having a responsibility for safety – railways and nuclear power for instance – than where it does not – roads. I have my doubts as to how much weight it is sensible to give to this by building different values into the formulae. Mostly what is at work here are different perceptions over what is likely to cause political embarrassment, rather than cool appraisal of differences in value or suffering. However, such diversity in valuation may be justified if it can be said that people would in general be prepared to pay more to avoid certain kinds of death, because of the pain to them or the suffering and cost caused to others.

If consistency were the sole objective, then it would not matter what value was chosen, provided it was used to test policy and investment across the board. But as stated before, most measures commonly do have benefits other than the reduction of fatalities. They may reduce injuries or damage, or have traffic or energy benefits, for example, outside safety altogether. If an effective balance is to be struck between investments with different mixes of these benefits, a defensible life-value must be used. Moreover, if one were, say, to take the £100 million value allegedly implied by the building of a pipe offshore from Sellafield into the Irish Sea, as a yardstick, one would logically be committed to such a diversion of resources into safety expenditure that it would have much the same effect as using an infinite value of life.

However, there are various methods used to give a more objective basis for the valuation of a human life (see Marin 1992). They range from the old Ministry of Transport method, which was based on the value of production figure, which at one time had the bizarre consequence that any woman was better off dead from society's point of view, since in those days most women never worked for money and therefore their loss could not be measured in terms of output figures; or based on the value of consumption forgone; or by more economically defensible but still controversial ways of trying to estimate what people or society might be ready to pay to avoid accidents. One can experiment to try to establish values, or engage in survey research, although there is always the difficulty of avoiding the myopia that leads people to set a much lower value on accidents beforehand than afterwards. Although important and fascinating, I believe that there is much to be done first through consistency tests of the kind I have suggested, before one needs venture into these difficult areas.

Conclusion

In conclusion, therefore, the initial task depends on choosing a plausible value both to check consistency and to give some indication of the areas where returns from investment seem high and others where we should raise the question that the proposed level of investment may be excessive, bearing in mind that the resulting expenditure may be very great. In the worst cases, excessive investment in safety may reduce the competitiveness of British industry by comparison with its overseas competition, to the extent that is both unfair and unreasonable.

CHAPTER SEVEN

Participation in risk management decisions

TO WHAT EXTENT IS RISK MANAGEMENT BEST LEFT TO EXPERTS?

A sixth area of debate concerns the optimum size and composition of the groups involved in making decisions on risk management issues. In other words, should as wide a range of people as possible have access to information on risks and be involved in management decision-making, or are there certain categories of risk that should be solely the domain of the "expert".

In one camp are those who are critical of narrow "technocrat" participation in decisions on sensitive issues, such as nuclear waste transportation (cf. Kirby 1988), and advocate extension of the "peer communities" who have, traditionally, been involved in risk management. The case for extension can be put in several ways. One is that opening up the relevant decisions and monitoring processes to wider scrutiny and attention from the multiple stakeholders involved will result in better-informed and less error-prone decisions. For example, the RCEP report (1989: 47, para. 6.39) noted that "The discovery of the environmental effects of DDT, leading eventually to its banning, is attributed to amateur ornithologists who noticed the decline in populations of peregrine falcons and other birds of prey" and went on to propose a broad basis of environmental monitoring activities. Extension of participation may also be argued to bring different scientific perspectives to bear. For example, some social scientists argue (see Sime 1985) that engineering-based crowd safety designs that treat human movements as analogous to the motion of physical objects will fail to model the essential characteristics of human crowds as interactive communication systems, and hence that broader participation in crowd safety decisions will avoid the serious errors that would otherwise arise.

Apart from its claimed effects in improving the information base of risk management, the case for broader participation is also sometimes put in terms of increasing the accountability of the technical decision-makers (Beder 1991), or on moral or "spillover" grounds. OECD guidelines on the management of chemical plants suggest that risk management decisions should explicitly consider suppliers, clients, customers and local residents, as well as corporate managers and employees.

The case for broader participationism is often linked to the development of challenges to the traditional positivistic view of scientific knowledge in the mould of eighteenth-century physics. Such challenges have developed in analyses of science (Wynne 1992c), of decisions about technology (Latour 1987, Collingridge 1980), and of the operation of the regulatory process (Stigler 1988). Funtowicz & Ravetz (1990, 1991, 1992) have argued that in circumstances such as global warming – where facts are inherently uncertain, values in dispute, stakes could be high and policy decisions presumed to be urgent – the institutional characteristics of "normal science" need to be significantly modified. They claim that in cases where neither conclusive scientific proof nor effective technology can be expected within the critical time-horizon of the decision, quality assurance of the (uncertain) scientific inputs to the policy process requires an "extended peer community". A short outline of their argument for a "post-normal" science approach to risk management can be found later in this chapter (see pp. 172–82).

The extension of participation is already accepted in practice where the ethical complexities of scientific work cannot be resolved within the orthodox boundaries of science and where non-scientists representing special perspectives and interests, set permissible limits on scientific work; embryo research is a most notable example. Similarly, in epidemiology there is claimed to be increasing participation of citizens in the identification of new medical problems from local and anecdotal data (Brown 1987).

In "trans-scientific" settings of this kind, the extension of participation in decision-making is not, according to Funtowicz & Ravetz, prompted by benevolence. It is a functional necessity for improving the quality of both decision-making and implementation, by broadening the base, first of knowledge and criticism, and then of consensus and responsibility. In this way, they claim, a new "social contract of science" can be achieved, in which there is a common respect for a plurality of competences, perspectives and commitments among the different stakeholders in a risk management issue.

Against this view are ranged those authors, such as Yalow (1985), who are sceptical of the benefits to be gained from broader participation in risk management decision-making and argue instead for the continuation of those decision-making methods that involve a few scientifically well informed participants, and which are still the most widely used in many areas of risk management. Those of this persuasion hold that proper risk management decisions require the application of the best available technical expertise to the reaching of consensus on the balance of the evidence. Yalow sets her argument in the context of what she sees as unfounded public fears about radio-immunoassay, and holds that the extension of participation in decision and policy-making processes may lead to quality scientific expertise being overridden by ill informed contributors or "junk science", so that risk management would become both impoverished and irrational as a result of whipped-up scares and over-politicization. In addition, it can be claimed that broad participation may be counterproductive in achieving its aims. If the only people who properly understand the risks of a project are those who are actively involved in its development, then broad participation will diffuse responsibility away from those who essentially make the decisions, thereby making it possible for them to lay the blame on poorly informed participants should things go wrong.

These themes are examined further in the three essays that follow. First, Nick Pidgeon examines arguments for narrow participation in risk management and comes to the conclusion that better results might be achieved by wider participation, while maintaining that there are exceptions and limits to the extent to which risk decision-making can be opened out to general debate. Then Silvio Funtowicz and Jerome Ravetz argue that we are now in the era of "post-normal" science as far as risk management is concerned, an era in which scientific evidence has to be seen to be intimately related to societal and ethical issues. A natural response to this development, they conclude, is the need for broader participation of stakeholders in decision-making. Finally, Timothy O'Riordan pursues a similar line of reasoning to argue for the growth of civic science, a phrase used to emphasize the point that the management of complex problems, such as risk, should be a participatory process.

TECHNOCRACY, DEMOCRACY, SECRECY AND ERROR

Nick Pidgeon[1]

Introduction: a sceptic's view?

Any scientific enterprise that aims to influence policy will inevitably have its political dimension. Decisions about risks – to health and safety, to the environment, or to the social fabric – are no different. They transcend, and therefore cannot be restricted to, such apparently neutral scientific questions as what are the uncertainties and consequences of hazards, and what methods and standards for assessment of risk should be adopted, but instead become a part of a much wider political discourse. This point is illustrated no more clearly than in the debate surrounding "narrow" or "broad" participation in institutional risk management.

In practice, the meanings of the terms "broad" and "narrow" are contested matters in and of themselves. Here I take "narrow participation" to refer to society entrusting issues of assessment of risk, as well as decisions about what risks to tolerate, primarily to a closed and elite circle of scientific and policy-making experts. In the narrow participation model the wider public (in theory at least) does hold some influence over the processes of risk decision-making, through the normal functioning of the political system. Such a state of affairs, as is the case in the Western democracies, generally represents the *status quo*. Fiorino (1990: 226) characterizes the case for narrow participation as follows:

> Many observers argue that risk decisions are best left to administrative officials in concert with scientific experts, acting under instructions from elected representatives, and consulting as necessary with interest groups representing aggregated public interests. Given the sheer complexity of the issues, the "trans-scientific" nature of the factual premises, and the rapid changes in the definition of problems and their solutions, the lay public lacks the time, information, and inclination to take part in technically based problem solving. Elites, it is argued, will make more rational decisions.

In recent years, however, several arguments have emerged for more broad forms of participation in risk management. Fiorino (ibid.) argues

---

1. I wish to acknowledge the helpful comments of Adrian Cohen, Christopher Hood and Tom Horlick-Jones.

that the ordinary citizen does have a more important role to play in the processes of risk decision-making and management, over and above that assumed in the narrow model outlined above. The belief that the public must be encouraged and enabled to become more actively involved in this way has found expression in the adoption of right-to-know legislation in the USA and elsewhere (Hadden 1989, Baram 1991), and in certain forms of risk communication that genuinely aim to empower, rather than merely to preach to, ordinary people as part of a genuine dialogue between the risk producers and risk bearers in society (see National Research Council 1989, O'Riordan 1990, Pidgeon et al. 1992). For example, in her analysis of the impact of right-to-know legislation in the USA, Hadden (1989) lists four possible participatory functions that it might serve: to enable citizens to find out about the risks they face; to contribute towards risk reduction (through improved emergency planning or changed behaviour of the regulated); to allow greater participation in societal decision-making; and, finally, to empower citizens in relation to corporate and government interests.

The present section takes as its starting point the views, first, that questions of risk, and in particular those of acceptability or tolerability, are at root decision-making problems (Fischhoff et al. 1981) and, secondly, that on an *a priori* basis, wide participation in societal decision-making is a desirable goal in and of itself. In this sense, the contribution is written from an initial position of scepticism regarding the sustainability of the narrow participation position. However, by exploring the three common strands of argument surrounding the narrow/broad participation debate – normative, instrumental and substantive – the aim is to ask under what circumstances arguments for narrow participation might or might not be sustained? The complexities of the institutional risk management debate mean that these strands are related to one another in complex ways, and as a consequence we should not necessarily expect to find simple, or universal, solutions to our common problems.

Three strands of debate: normative, instrumental and substantive

The normative argument: technocracy and democracy
On strictly normative grounds, and to this author at least, technocracy must always be subordinate to democracy (see also Stern 1991). Such a belief can be justified either by reference to political realism (the assessment that in certain social and legal systems, such as in the USA or certain countries of the European Union, the public will force its way into the process anyway) or to political idealism, such that more rather than less

participation in the policy process is an end to be valued and worked for in and of itself (Fischhoff et al. 1983). Probably few would publicly deny this general proposition.

However, the case for narrow participation in risk management is sometimes justified on the grounds that science is an ethically agnostic activity and that, under such circumstances, entrusting risk decisions to scientists is not subject, in theory at least, to the vagaries of the political agenda or to passing fashion and fad. Under these circumstances, long-term planning and the emergence of best (or at least the avoidance of worst) decision options is facilitated, through the scientific discovery of "truth" and the facts of the matter unbiased by hidden assumptions, value conflicts or the political agenda.

The Enlightenment assumptions inherent in this view, although convenient and familiar to many scientists schooled in both the social and the natural sciences, may ultimately prove unsustainable. A first point is that to argue in this way is self-contradictory, since this position is not in itself ethically neutral (implying, as it does, that democracy can be conveniently subordinated to technocracy). A second, more fundamental, philosophical point is that science itself, as an essentially human and social activity, is not value neutral in either practical or epistemological terms (e.g. Ravetz 1971, Latour 1987). For example, most risk professionals work within institutional contexts, such as large corporations and their associated private sector consultancies, government regulatory agencies, or environmental groups. As Dietz & Rycroft (1987) report, these affiliations (particularly those of the corporate variety) have more than just a passing relationship with individual scientists' research agendas and value orientations. Set against this is the counter-argument that the cumulative impact of the scientific process, and with it sustained contact with the empirical world, does eventually bring progress and truth through the identification and elimination of hidden agendas and social biases (see Gross & Levitt 1994).

A final point is that many of the hotly contested conflicts generated by risk issues are themselves at root questions of value (von Winterfeldt & Edwards 1984); for example, regarding fairness and the distribution of harms across society, tolerability, or the importance placed by different groups upon particular harms and consequences. Hence, the acceptability of risk is *inherently* political (Douglas 1985). To take but one example, the argument, implicit in many risk evaluation approaches, that society should maximize lives saved for a given investment, assumes that economic efficiency in reducing fatalities is the universal value (Rayner 1989). In this respect, therefore, scientific arguments can only contribute one element, albeit a crucial one, to the total risk policy process. As Shrader-Frechette notes "to attempt to reduce [risk questions] to purely scientific issues, is to ignore the value dimension of policy analysis and to disenfranchise the public who, in a democracy, ought to control that policy" (1985: 151). This is probably the *strongest* normative argument for respecting broad forms of participation in risk decision-making, although it does not of itself resolve the rather more tricky question of how society's value priorities and sound scientific knowledge might be best *combined* in risk decision-making.

The instrumental argument: legitimation and liability
Traditionally, the narrow participation model is justified (and is seen as legitimated) on the grounds that decisions that are taken in secret will nevertheless be sensitive to citizens' interests and agendas, and that over the long run we can trust existing institutions broadly to reflect the interests of all. For example, in the UK, current (mid-1990s) health and safety regulatory practice (see HSE 1992a), reflecting earlier psychometric work on expressed risk preferences, distinguishes between *individual risks* (which arise from exposure to events involving single or few fatalities, such as motor-cycle or automobile accidents) and *societal risks* (potential large-scale loss of life from a single event, such as a civilian aircraft disaster). Although there are several unresolved definitional difficulties associated with the societal risk concept, the wider implication is that safety standards, and with it the price to be paid for safety controls, might sometimes need to be set differently across such different risk contexts. Such a course, which reflects public concerns over such things as the involuntariness and uncontrollability of large-scale accidents, is not necessarily inconsistent. From the risk manager's perspective, when the full range of consequences are totalled, societal risks often do have wider and more varied impacts – such as loss of consumer confidence and sales through social amplification effects, stricter regulation, and in some cases bankruptcy – than that of the direct threat to life and limb alone that individual risks pose to workers or isolated members of the public. Hence, it could be argued that here, at least, the *status quo* already does, if only imperfectly, reflect certain wider aspects of society's views on the quality of risk in the risk management and regulation process.

Set against this is the argument that traditional political institutions are no longer sufficient to deal with the changed hazards (and the social contexts of those hazards) of the late-modern world, creating a crisis of legitimation. Here some would argue that control over ever more complex hazards has increasingly become invested in large-scale institutions, in ways that may be remote from citizen (or even politicians') influence. At the same time, and for a variety of reasons, citizens may have become less willing to trust such institutions to act responsibly and fairly upon their behalf (see Laird 1989, for some circumstantial evidence on this point). It may be significant here that the finding, noted above, that perceived control is a critical aspect of how people construe hazards, may be indirectly reflecting the wider institutional question of the trust relationship between producers and bearers of risk.

If we accept the view that institutional trust is under threat, and with it the implication that there is a crisis of legitimacy in risk management, this does indeed set us a challenge to invent new decision forums. Calls for broad participation can therefore be seen as a symptom, *if not necessarily the only potential solution,* to this problem. Whether risk communication, right-to-know legislation, referenda, or citizens evaluation panels (Shrader-Frechette 1985) will indeed bring such legitimacy, is a matter for further reflection and research, particularly regarding both the strengths and the practical limitations of possible mechanisms (see Fiorino 1990). However, two arguments do suggest that the search for wider legitimacy in risk management processes may not be as easy as it might at first seem.

A first, pragmatic argument, is that greater (genuine) participation might indeed bring more legitimate *processes*, but at the expense of a much wider diffusion of *responsibility* for decisions. Under some circumstances this might adversely affect safety: for example, where responsibility shared is responsibility lost (Turner 1978). Diffusion of responsibility also raises the question of liability if things subsequently do go wrong. One unintended consequence of this for society, might be that subsequent prosecutions of corporations and individuals that act negligently might be made more difficult, as responsibility for failures shifts from risk producers to other involved groups (and particularly to a society's regulators). Other consequences of diffusion of responsibility may be that decision-making will occur through default or delay (a decision to delay is still a decision!), or that the locus of true decision-making still remains secret, and has merely been shifted elsewhere by powerful institutional interests. A final possible drawback of openness, where inherently uncertain premises and judgements can be legally challenged after the event with all the benefits hindsight brings, might be a move to more defensive (but legally defensible) forms of risk management, where discretionary case-by-case (and hence arguably more flexible and resilient) approaches are replaced by more rule-based (anticipatory) forms of risk management (see Rimington 1993). Such a move might be desirable only in contexts where anticipatory risk management can be shown to hold a clear advantage over the promotion of resilience (see also Collingridge, Ch. 2, this volume).

A second, more theoretical point, flows from research findings on risk perceptions, which show that social and cultural factors are critical determinants of perceived risk, and that different groups in society will exhibit fundamentally different (and legitimate) perspectives on such matters. Irrespective of whether these are framed in terms of attitudes, environmental world-views, or the defence of institutional forms and cultural biases (Thompson et al. 1990), we have undoubtedly now entered the age of "plural rationalities" in risk matters. The call for broad participation is of course compatible with, and arose in part from, the plural rationalities perspective. As Kemp (1993: 112–13) rightly notes, however:

> . . . cultural theories of risk perception are descriptive rather than prescriptive, and they fail to provide ways out of the ensuing problem of relativism that they raise; hence, if scientific expertise continues to be questioned, and if competing views of nature are equally legitimate, how do we escape the implicit relativism which underpins cultural theory in order to resolve waste disposal and incinerator siting problems and the like?

Note that this is not an argument against using competing arguments and perspectives to uncover errors of judgement and fact (which is considered further below), but rather a more fundamental problem of whether we can ever resolve different, and possibly contradictory, value positions within stakeholder groups (even when it is clearly useful to the policy process to bring the fundamental causes of conflict to light)? In response, one can point to the potential for learning inherent in communication and political discourse around values. On the other hand, in some public forums where competing viewpoints are aired (such as courts of inquiry), actors with differing interests and values are inevitably placed in adversarial opposition, reducing the chances of genuine dialogue and debate. Research on the social psychology of intergroup relations (see Tajfel 1981) is not encouraging as to the possibilities for conflict resolution under such circumstances!

The substantive argument:
wider constituencies reveal hidden assumptions and error

The substantive question of whether more effective risk assessment and management is facilitated by narrow or broad participation is hotly contested. Some complex arguments are involved here. On the one hand, researchers such as Funtowicz & Ravetz (1990) argue that the complexities of many risk problems are such that systemic uncertainty and ignorance are endemic, values in dispute, and stakes high. In trans-scientific settings such as these, the institutional characteristics of "normal science, need to be modified, the extension of participation being necessary for improving the *quality* of both decision-making and implementation, by broadening the cognitive base of knowledge and criticism, as well as the social base of consensus and responsibility.

For example, if we consider probabilistic risk assessment, it is clear that widening the perspectives to be brought to bear when an analytic model is initially constructed is one important way to avoid blind spots in the subsequent analysis (Pidgeon 1988). Here we know that, by considering how human and organizational factors influence reliability in what at first sight might seem purely technological systems (Turner 1978, Blockley 1985, Pidgeon & O'Leary 1994, Toft & Reynolds 1994), certain forms of modelling incompleteness and systemic uncertainty can be countered.

A second, more speculative, point is that involving communities in risk communication programmes that allow for a genuine two-way dialogue and active learning by all stakeholders, leads, in Westrum's (1993) terms, to more "generative" institutional systems, and perhaps even to improved risk management practice through early hazard identification and correction. This suggestion can be justified by reference to sociological case studies of lay understandings of scientific controversies, which suggest that relevant knowledge does not always reside solely with formal scientific experts and risk managers. This point is also borne out by the fact that, when a closer look is taken at some of the supposed biases in lay evaluations of risk, such judgements often appear to be relatively well founded, and they reflect several relevant qualitative aspects of risk and hazard that cannot always be accommodated by formal risk models (see, for example, Fischhoff 1990). Added to this is the observation that experts themselves are not always immune from judgemental biases (Fischhoff & Svenson 1988), or from incorporating implicit framing assumptions in their risk analyses (Wynne 1992a).

However, the theoretical possibilities for error reduction brought by introducing more and plural voices into the process has to be balanced against the potential unintended consequences of introducing more "noise" to the risk management system (Turner 1978). Furthermore, there may always be some circumstances where secrecy has to be maintained, and therefore different, essentially closed, institutional mechanisms for error and risk reduction sought. Take, for example, the command and control of nuclear weapons systems. In his analysis of some of the historical organizational and operational failings of the US system of control, Sagan (1993) highlights how, for procedural and political reasons, the US command and warning organizations collectively failed to learn the lessons of several serious breakdowns of command and control (e.g. the hurried use of temporary, surveillance radars during the Cuban missile crisis, which were prone to false warnings). Sagan concludes, correctly, that one way, although not the only way, to improve learning will be to challenge the vested interests that stifle learning through independent civilian oversight of military command and control organizations and activities. However, for some fairly obvious reasons (threats of terrorism or foreign espionage among them), this does not mean that the veil of secrecy should be lifted and the technology exposed to full public view. Of course, one is then left with the question of how "cognitively open" an essentially closed system of risk management can become?

Concluding comments

This section has examined some of the complexities of the narrow/broad participation debate, and some of the institutional dilemmas that this raises. A first conclusion is that there are clear normative arguments for wider forums of participation in risk decision-making. All other things being equal, less secrecy is a goal to be valued in and of itself, and several legitimation considerations would support this conclusion. However, against this are set practical objections to broad participation (revolving primarily around issues of blame and responsibility), as well as difficulties raised by the recognition that social discourses about risks reflect plural rationalities. The former might perhaps be overcome by institutional adjustments, and are certainly issues for further reflection and research. The latter, however, sets us a more formidable philosophical conundrum. Finally, although substantive arguments do suggest that broad participation may lead to the detection of errors, and with this better chances for learning and improved risk management, there will also be circumstances in which it brings other, perhaps worse, risks in its train. Ultimately, the challenge remains to understand how the social contexts of risk shape appropriate institutional strategies for its management in a world beset with dilemmas.

RISK MANAGEMENT, POST-NORMAL SCIENCE, AND EXTENDED PEER COMMUNITIES

Silvio O. Funtowicz & Jerome R. Ravetz

Introduction

Few will still doubt that our modern technological culture has reached a turning point, and that it must change significantly if we are to manage problems of risk and the environment. It may not yet be widely appreciated that science, hitherto accepted as the mainspring of technological progress, must also change. These problems of risk and the environment present new tasks for science; along with the discovery and application of scientific facts, new fundamental achievements for science must also be concerned with remedying the pathologies of our industrial system. We no longer require the ideal of a science that is totally value-free and ethically neutral, nor do we need to believe that rational and correct policy decisions automatically follow from the facts discovered by science. A new method, based on the recognition of uncertainty, complexity and quality, will guide the new scientific enterprise, which we call "post-normal science".

Our approach is new in its emphasis on the concepts of uncertainty, complexity, and quality. All these had previously been kept at the margin of the understanding of science, among researchers, philosophers and popularizers alike. Science was traditionally imagined as steadily advancing our certain knowledge and effective control over the natural world. Now science is appreciated as confronting complexities and coping with uncertainties in urgent decisions on technological risks and environmental threats on both global and local scales. The work of quality assurance of the results of research in this new, broader context of science can no longer be left to isolated specialist communities; it must be renewed and enriched. The dialogue on quality, along with that on policy, must be extended to all those with a stake in an issue who are committed to a genuine debate; these we call "the extended peer community".

We have developed a method for assessing and expressing the quality of technical information in terms of its characteristic uncertainties (Funtowicz & Ravetz 1990). Called "NUSAP", an acronym for the names of the five boxes of a standard notation, it systematizes and generalizes good scientific practice. Along with the numerals and units of any quantitative expression, it has a third box for spread, which corresponds to the "error bars", which are an essential part of the information in any genuine scientific expression. In addition, it provides two further categories:

"assessment", a generalization of "accuracy" or "systematic error", codes for the quality of the information; "pedigree", in which relevant aspects of the mode of production of the information are displayed in a matrix. This provides the basis for the assessment of quality; and enables users of the information to make their own judgement at a glance, rather than being referred to lengthy small-print explanations or appendices. As it becomes diffused, the "NUSAP" system will facilitate the critical assessment of technical information, thereby contributing to the work of quality assurance in the extended peer communities that engage on problems of risks and the environment through post-normal science.

Post-normal science

The increasingly complex and urgent problems of risk and the environment have common features that distinguish them from traditional scientific problems. They vary in scale from the local to the planetary, and are often very long term in their impact. The phenomena are novel, complex and variable, and are often not well understood. Data on their effects, and data for baselines of "undisturbed" systems, are frequently inadequate. For these new problems, science cannot usually provide well founded theories, based on experiments, for explanation and prediction. Frequently it can achieve no more than mathematical models and computer simulations, neither capable of being tested by traditional scientific methods. On the basis of such uncertain inputs, policy decisions must be made, often under conditions of some urgency. Therefore, policies for solving the environmental problems cannot be determined on the basis of scientific predictions, but only supported by policy forecasts.

We adopt the term "post-normal" to mark the passing of an age when the norm for effective scientific practice could be a process of routine puzzle-solving (Kuhn 1962) conducted in ignorance of the wider methodological, societal and ethical issues raised by the activity and its products. The leading scientific problems can no longer derive solely from the curiosity of scientists or the missions of defence, industry or medicine. These new problems are created by issues where, typically, facts are uncertain, values in dispute, stakes high and decisions urgent. The community of researchers do not have the luxury of deferring investigation until they are hopeful of success; in the area under discussion here, researchers must do their best, however complex the problem and uncertain the solution. Moreover, when research is called for, there must first be a definition of the problem to be studied, and this will depend on which aspects of the issue are most salient. Hence, political considerations constrain the possibilities of the sorts of results that can be produced, and thereby the sorts of policy options for which there is scientific support. In general, the situation of post-normal science is one where the traditional opposition of "hard" facts and "soft" values is inverted; here we find decisions that are "hard" in every sense, for which the scientific inputs are irremediably "soft".

The inherent limitations of the traditional problem-solving strategies are revealed by a structural feature of the new problems of risk and the environment. For in these, decisions depend on assessments of future states of the "natural" environment, resources, technology and human society, all of which are currently unknown and also unknowable in any detail. Further, in addition to the irremediable uncertainties in knowledge relevant to policy, science-based technology has created moral complexities resulting from the invasion of the domains of the sacred and private. The most notable cases are reproductive technology and also scientific research that requires the inflicting of harm on aware beings. Under these new circumstances of radical uncertainties of every sort, a new type of problem-solving strategy is emerging.

We can analyze the different sorts of problem-solving strategies that are now employed, through a biaxial diagram (Fig. 7.1), which exhibits them with reference to the two attributes of "systems uncertainties" and "decision stakes" (Funtowicz & Ravetz 1992). For systems uncertainties, the three intervals along the axis correspond to different sorts of uncertainty, namely technical, methodological and epistemological. The other axis relates practice to the world of policy. For decision stakes, we understand in general the costs, benefits, and commitments of any kind, for the various stakeholders in an issue. There are three zones, corresponding to three types of problem-solving strategy: applied science, professional consultancy and post-normal science. (Traditional "pure" science would, on this diagram, be located at the intersection of the axes.)

There is no pretence of quantifying either of the attributes defining the problem-solving strategies. They provide a rough gauge whereby the distinctions among the three zones can be illuminated. When both attributes are minimal, then routine puzzle-solving research in the Kuhnian sense is adequate; this occurs when the research contributes a useful piece of information that is neither contested nor critical in relation to a policy issue. But when either attribute is medium, something extra must be brought into the work, which can be called the professional's skill or judgement. For professional consultancy, the attributes may range from moderate to severe; the medical doctor normally cares for the health or life of individual patients, although the task is more demanding in times of public health crises; whereas for the engineer there is the welfare of a client and, in connection with safety, that of a wider community. In post-normal science, when problems of risk and the environment are involved, the stakes can become the survival of a civilization or ecosystem or, for

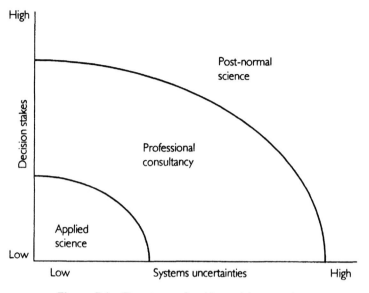

Figure 7.1   Three types of problem-solving strategies.

example, present forms of life on the planet; and the systems uncertainties are correspondingly severe.

The diagram displays the feature that, even when uncertainties are low, if decision stakes are high then "applied science" puzzle-solving alone will not be effective in a decision process. For no scientific argument can be logically conclusive; even the traditional positivistic philosophy of science acknowledged this. In the course of a scientific debate, the arguments evolve in a continuous dialogue that is incapable of reduction to logic; what makes scientists "rationally" change their opinions is a matter of continuing discussion among philosophers and sociologists of science (Chalmers 1990). Applying this lesson to debates on particular issues (as in the regulatory system), we can appreciate that when any party finds its interests threatened it can always identify some methodological weakness by which to challenge the quality of the scientific information presented by the other side. This is particularly easy in the case of regulatory decisions on risks or the environment, where the uncertainties of evidence and argument are severe. Thus, in the policy arena, the forum for scientific debate becomes enlarged from that of the technical experts alone, to include all those interests, commercial or corporate, with a strong stake in the outcome.

All these tendencies to a broadened forum of debate appear still more strongly in the case of post-normal science. Research work and the deployment of skills still have an essential role to play, but this must be done within a framework in which the narrowly defined scientific problems are

175

integrated into larger policy issues. In this way they are provided with direction, quality assurance, and also the means for a consensual solution of policy problems, in spite of their inherent complexities and uncertainties. Examples of problems with combined high decision stakes and high systems uncertainties are familiar. Any of the problems of major technological hazards or large-scale pollution belong here. The paradigm case for post-normal science could be the design of a repository for long-lived nuclear waste, required to be secure for the next 10 000 years.

The usefulness of our diagrammatic scheme can be illustrated by consideration of cases located close to either of the axes. For a problem with low systems uncertainties, we have examples among the major disasters that have afflicted our modern industrial societies in recent years. Subsequent inquiries have, in many cases, established that the disaster had been "waiting to happen" through a combination of physical predisposing causes and management practices that had been well known in advance (e.g. Bhopal, *Challenger*, *Exxon Valdez*). Yet applied science and professional consultancy were not enough to prevent the accidents in the first place; and the strengthening of the regulations for avoiding recurrences requires that disasters become policy issues, to be eventually resolved through post-normal science.

Cosmology is a contrasting case, a science that now (unlike in Galileo's time) has low decision stakes along with high uncertainty. Here the data are so sparse, theories so difficult to test, and public interest so lively, that the field is as much "natural philosophy" as science; and experts must share the platform with amateurs, popularizers, philosophers and even theologians. In this example we see an historical continuity between the science that was practised before the establishment of the authoritarian paradigms, and the emerging post-normal science of the present. This can help us appreciate the methodological continuity between post-normal science and all other problem-solving strategies. For post-normal science is a development of traditional forms of science, one that is appropriate to the conditions of the present age. Its essential principle is that, in science-based policy decisions and even in science, we can no longer expect to conquer or banish uncertainty and ignorance. Instead, they must be managed for the common good. Programmes for the reform of technology, industry or life-style that ignore this aspect of contemporary scientific knowledge are likely to remain part of the global problem rather than to contribute to its solution.

By the use of the diagram (Fig. 7.1), we can better understand the different aspects of complex projects in which all three sorts of practice may be involved. For this we may take an example of a dam, as was discussed previously (Ravetz 1971) in connection with an analogous classification of problems as scientific, technical and practical. First, in the construction of a dam there is use of basic, accepted scientific knowledge; and there will be particular research projects of an "applied science" character to provide information on the relevant features of the local environment and details of the dam's construction. But the creation of the dam is also a design exercise, where the shape and structure is not completely determined by scientific inputs. If nothing else, there will be a design compromise among the various possible functions of the completed dam, which may include water storage, hydroelectric power, flood control, irrigation and leisure, together with their associated costs. Achieving an optimum balance among these, given both the uncertainties in scientific inputs and the value-conflicts among the various affected interests, is a task for a "professional consultancy". But the matter does not stop there. There may be a possibility (although no certainty) of long-term change in the hydrological parameters of the catchment, of adverse effects down stream, and perhaps even local earthquakes. Some people may find their homes, farms and religious monuments drowned by the artificial lake; can they possibly be adequately recompensed? Dams, once seen as a completely benign instrument of human control over raw nature, have suddenly become seen as a sort of predatory centralism, practised by vast impersonal bureaucracies against local communities and the natural environment. When such issues are in play, we are definitely beyond professional consultancy, and we are in the realm of post-normal science. Also, we observe that the "complexity" of the dam project does not lie essentially in the variety of relevant scientific disciplines, but rather that it consists of the multiplicity of legitimate perspectives on the total issue.

Extended peer communities

We can also use the diagram to illustrate how a problem in post-normal science can evolve so that it is brought some way in towards manageability. When, for example a risk or pollution problem is first announced, it will almost always be in a condition of considerable uncertainty. Since it had not been appreciated previously, there is hardly likely to be substantial information about it. Hence, the evidence will tend to be anecdotal on the experimental side and speculative on the theoretical side. But the strength of the decision stakes will ensure that all interests will offer their opinions with apparently complete certainty. The first phase of the discussion will, therefore, resemble ordinary political debate, but of a particularly confused kind. Each side will attempt to define the problem in the terms most favourable to its interest; typically proponents of a development presenting it as straightforward applied science and opponents stressing uncertainties and ethical aspects. It is a new phenomenon for such broad debates to be effective; hitherto, commercial viability or state security was the overriding consideration for industrial development, subject to some concern for health, safety and the environment. Indeed, in recent decades, traditionally trained experts have experienced bewilderment and dismay as they confronted those who try to block "progress" on the basis of apparently intangible and non-scientific arguments.

If such problems remained in the realm of pure power-politics, the outlook for our policies for science, technology and the environment would be grim. But there is a pattern of evolution of problems, with different problem-solving strategies coming to prominence. This gives hope that professional consultancy and also applied science may yet have important roles to play. For as the debate develops from its initial confused phase, positions are clarified and new research is stimulated. Although the definition of problems is (as we have seen) never free of politics, an open dialogue can ensure that such considerations are neither one-sided nor covert. In the developing discussion on the technical aspects, no advocates need admit they were wrong; it is sufficient for there to be a tacit shifting in the terms of the debate. And, as new research eventually brings in new information, the problem becomes more amenable to the approach of professional consultancy and, for example, of applied science. Thus, by means of Figure 7.1, we can indicate a pattern for the progressive evolution of a complex and uncertain issue involving science and policy.

It is important to appreciate that post-normal science is complementary to applied science and professional consultancy. It is not a replacement to traditional forms of science, nor does it contest the claims to reliable knowledge or certified expertise that are made on behalf of science in its legitimate contexts. Recent critical philosophies of science, concentrating on scientific knowledge alienated from its social context, have led to a view that "anything goes" in science. It is as if any charlatan and crank should have equal standing with qualified scientists or professionals (see notably Feyerabend 1975). Our critical analysis proceeds on another basis, that of quality assurance, or critical assessment. The technical expertise of qualified scientists and professionals in accepted spheres of work is not being contested; what can be questioned is the quality of that work, especially in respect of its environmental, societal and ethical aspects. Previously the ruling assumption was that these were "externalities" to the work of science itself; and that, when such problems arose, an appropriate response would somehow be invented by "society". Now the task is to see what sorts of changes in the practice of science, and in its institutions, will be entailed by the recognition of uncertainty, complexity and quality within policy-relevant research.

In what we might now call "pre-normal" science, nearly all the practitioners were amateurs. They could and did debate vigorously on all aspects of the work, from data to methodology, but usually there was no in-group of established experts in conflict with an out-group of critics. In normal science, any outsiders were effectively excluded from dialogue; they would get a chance to be heard only in a Kuhnian "pre-revolutionary" situation, when the ruling paradigm (cognitive and social) could not deliver the goods in the way of steady puzzle-solving progress. In post-normal science there is still a distinction between insiders and outsiders, based (on the side of knowledge) on certified expertise and (on the social side) by occupation. But since the insiders are often incapable of providing conclusive solutions to the complex problems they confront, the outsiders are capable of forcing their way into a dialogue. When the debate is conducted before a lay public, the outsiders (including community members, environmental activists, lawyers, legislators and journalists) may on occasion even influence the agenda. An example is in biomedical science, where non-professional groups advise on ethical issues, and where activists have now joined the dialogue about the treatment of, and even research into, some of the more controversial diseases such as AIDS (Brown 1993a).

Because of these human aspects of the issues giving rise to post-normal science, there must be an extension of all the elements of the scientific enterprise. First, there must be a presence of a complementary expertise whose roots and affiliations lie outside that of those involved in creating or officially regulating the problem. These new participants, enriching the traditional peer communities and creating what might be called "extended peer communities", are necessary for the transmission of skills and for quality assurance of results. It is important to realize that this phenomenon is not merely the result of the external political pressures on science that occur when the general public is concerned about an environmental issue. Rather, in the conditions of post-normal science, the essential functions of quality assurance and critical assessment can no longer be completely performed by a restricted corps of insiders.

When problems lack neat solutions, when environmental and ethical aspects of the issues are prominent, when the phenomena themselves are ambiguous, and when all research techniques are open to methodological criticism, then the debates on quality are not enhanced by the exclusion of all but the specialist researchers and official experts. The extension of the peer community is then not merely an ethical or political act; it can positively enrich the processes of scientific investigation. Knowledge of local conditions may determine which data are strong and relevant, and can also help to defuse the policy problems. Such local, personal knowledge does not come naturally to the subject-specialty experts whose training and employment predispose them to adopt abstract, generalized conceptions regarding the genuineness of problems and the relevance of information. Those whose lives and livelihoods depend on the solution of the problems will have a keen awareness of how the general principles are realized in their "back yards". They will also have "extended facts", including anecdotes, informal surveys, and official information published by unofficial means. It may be argued that they lack theoretical knowledge and are biased by self-interest; but it can equally well be argued that the experts lack practical knowledge and have their own unselfconscious forms of bias. Indeed, since the investigation of a local problem may require governmental action, and this can be subject to a variety of pressures, so on occasion the barriers to the undertaking of scientific research will be overcome only as a consequence of pressure from concerned citizens (Ozonoff 1993).

An excellent example of how the normal practice of science can be changed by citizens' involvement is provided by the history of the Leukaemia cluster at Woburn, Massachusetts. At the beginning of their campaign in the 1970s, citizens encountered suspicion and hostility from both state agencies and established scientists. To them this may well have appeared as prejudice and obstruction, on behalf of the various vested interests in the case; but more likely it was attributable, at least in part, to the inexperience of the other side, with initiatives coming from the grassroots. By the end of the struggle, which took place on the political and scientific fronts simultaneously, "popular epidemiology" was an established and respected form of research, and the established institutions themselves accepted a new way of working (Brown 1993b).

This example shows that there is no question of saying whether it is the restricted or the extended peer community that has the "better" knowledge. Rather, we should see them as complementary, mutually supporting and reinforcing. Indeed, with the perspective of this sort of practice, we can envisage a new, humanistic goal for science and technology. In postnormal science, we weaken the logical ideal of "scientific prediction", and are satisfied with the more pragmatic goal of "policy forecasting". However, in regard to the knowledge gained, we can enhance the traditional conception of "scientific explanation" to a richer "societal understanding". In this way the new challenges and the emerging practice of postnormal science can lead to new, appropriate ideals for science itself.

Conclusion

Technologically advanced societies have now reached the point where the traditional strategies of scientific problem-solving are no longer appropriate to new needs. Unless we find a way of enriching our research endeavour to include this new sort of practice, we will fail to develop methods for meeting the new environmental challenges, with all their complexity and uncertainty.

Fortunately, the conditions are ripe, in the broadening social distribution of knowledge and skills. In modern societies, including some of the poor as well as the rich, there are now large constituencies of ordinary people who can read, write, vote and debate. The democratization of political life is now commonplace; its hazards are accepted as a small price to pay for its benefits. Now it is becoming possible to achieve a parallel democratization of knowledge, not merely in mass institutional education but also in enhanced participation in decision-making for the wise management of our scientific powers.

The democratization of this aspect of science is, therefore, not a matter of benevolence by the established groups, but (as in the sphere of politics) the achievement of a system that, in spite of its inefficiencies, is the most effective means for avoiding the disasters that result from the prolonged stifling of criticism. Recent experience has shown that such a critical presence is as important for our technological and environmental issues as it is for society. Let us be quite clear on this: we are not arguing for the democratization of science on the basis of a generalized wish for the greatest possible extension of democracy in society. The epistemological analysis of post-normal science, rooted in the practical tasks of quality assurance, shows that such an extension of peer-communities, with the corresponding extension of facts, is necessary for the effectiveness of science in meeting the new challenges.

EXPLORING THE ROLE OF CIVIC SCIENCE IN RISK MANAGEMENT

Timothy O'Riordan

Science in flux

It is hard to put a finger on it, but one senses an uneasiness in the world of science. Nothing is particularly noticeable, but there are straws in the wind.

The so-called "new physics" and "new biology" are undermining long-established theories about the properties of matter, energy, evolution, competition and cooperation. There is a vast literature on this, but a start can be found in Bohm (1980), Davies (1987) and Russell (1993). It is just possible that out of this flux of half-proven ideas will come a coherent language of communicative intelligence and self-organization that will suggest a "meta-science" of physical, chemical and social interaction through which learning and adaptation are signalled across energy fields and time in ways that are meaningful yet mysterious.

Discipline-bound science, rooted in the epistemology of positivism, verification and bounded peer review, is finding it difficult to be credible in the emerging participatory worlds of risk management and global environmental change (O'Riordan & Rayner 1992).

Social science methodologies that seek generalizations on individual heuristics of response to uncertainty and anxiety, expose a failure to recognize the overwhelming significance of inner world views and the socialization of experience (Eiser 1994).

"Expertise" is becoming devalued. This is partly because there are always more "experts" than a problem can handle. But more to the point is a growing disillusionment that "experts" can truly speak for the "public good". Sufficient numbers of people now look for expertise to be both a participatory experience and a genuine dialogue of equivalent power and justice (Renn & Levine 1992)

The geography of environmental issues and deprivation mirrors the geography of poverty, social disorganization and political powerlessness. There is a feeling that risk management may actually have reinforced the inequality of environmental opportunity rather than diminished it (Bullard 1994).

Uncertainty is coming to mean indeterminacy, namely the inability to understand process and outcome, and hence to predict futures based on simplifications of present states and their dynamics. This means that traditional ways of comparing and judging the merits of many possible outcomes cannot be achieved by "pure" rationality and logical analysis. Judgement, based on honest and fair debate, is needed, and science lacks the mechanisms to do this. Cost–benefit analysis is becoming discredited for its distortion of "true" values (see O'Neill 1993 for a robust statement), and courts of law become discomforted when experts cannot command with unambiguous authority when cause and effect are in flux.

These are hunches. Much of the science world of grant-seeking, networking and jet-setting remains largely oblivious of these stirrings of peripheral turmoil. The dominant powers in the world of science are dismissive of the "new" physics and biology, resistant to true interdisciplinarity, and unwilling to regard science as an outcome of participatory negotiation with citizens' groups and community activists. Although Grove-White & Shiva (1992) would not agree completely with this statement, most of this introduction is influenced by their timely and pertinent analysis (also, see Aronowitz's (1988) clever critique of "old" science).

For the majority of its practitioners, science must remain true to its traditions, its methods, and its social role, if it is to remain purposeful, and their role worthwhile. Evidence must be supported by observation, replication, verification and hypothesis testing. Where there is doubt, internal peer review should determine what is acceptable and what is not. If peer review cannot resolve the matter, then the differing views should be published for all to see and to judge. Science should be seen as illuminating, enlightening, revealing and guiding. It is to serve the public interest by standing apart and providing a perspective. How the evidence is subsequently interpreted and utilized in political decisions is "beyond science"; that is claimed to be the proper job of politics. These distinctions may be a little blurred nowadays, but they still hold for many. But in the areas of risk management, global environmental change and the so-called "sustainability transition", these views are fundamentally challenged.

The rise of civic science

The phrase civic science is used by Kai Lee (1993) to emphasize the point that managing complex systems should be a participatory process, open to learning from errors and profiting from successes. Indeed, Lee makes an even more fundamental point when considering decisions over complex natural systems, such as river basin management. "Policies to learn", he comments (p. 161), "must persist for times of biological significance, and they must affect human action on the scale of ecosystems". In short, the institutions of decision have to be natural, adaptive and organic. These are not the patterns commonly found in economics, politics and the law. Civic science, therefore, poses an extension not just of science, but of the forms of choice-making in a modern democracy.

Thus, civic science politicizes science in the sense that science evolves via adaptive human choices about means and ends. This in turn throws the spotlight on the distribution of power, for adaptive human choices are overwhelmingly influenced by the structural and financial ability to be respected and responded to in open political debate. Piller (1991) provides a useful review of this point. Civic science is also inherently political in that society becomes more of a true partner in political decisions. Adaptive learning also means adaptive listening. This in turn requires structural arrangements that give the "civic" part of civic science a full voice in the evolutionary process of social learning.

Civic science has become recognized partly because of the sterling work by risk management researchers and analysts since the mid-1980s. The final two chapters of the Royal Society Study Group report on risk (Royal Society 1992: 89–192) cover much of this ground admirably.

This corpus of research has revealed a host of fascinating themes:

- *Risk is a culturally framed concept* that acts as a metaphor for individual feelings about control, powerlessness and the drift of social change in terms of good and bad for self and family. Risk is therefore multidimensional, not separable into mechanistic structures of probability and evaluation. Risk is also reflective of the social order of justice and opportunity. As the powerless become collectively more aware of their position, so they seek to use the metaphor of risk as a weapon of frustration, not just protest.

  This is why the so-called NIMBY ("not in my back yard") protest has broadened into a more fundamental critique over social injustice. Many, if not all, NIMBY groups are arguing more on generic or principled grounds than on a fight against an unwanted project in their midst. Hence, the opposition to toxic waste incinerators becomes a demand for "a greater reduction at source" approach to waste management (see Blowers et al. 1992). In a similar vein, the growing dislike of wind farms in rural areas of the UK becomes a matter of disaffection against excessive subsidies for renewable energy in the wake of underfunding for energy conservation measures in the regional electricity companies.

  Schrader-Frechette (1991: 53–66) provides an illuminating analysis of the significance of scientific rationality in risk management (as also does O'Neill (1993: 145–68). She points out that the cultural critique goes too far in removing the authoritative basis of scientific discourse. Hence, the need for a civic science that fuses the rational logic of the scientific tradition with the judgemental biases of democratic procedures.

- *Risk tolerance is quite different from risk acceptance.* There may be no such thing as risk acceptance where the nature of the threat is suspected or distrusted. Risk tolerance is managed by the delivery of secure arrangements for participatory involvement, compensation guarantees in the event of error or accident, and ultimate power over the future path of hazardous technologies or processes. The distinction was made in the Sizewell Inquiry over a major new nuclear reactor programme for the UK (see O'Riordan et al. 1987: 192–9). Curiously, but predictably, the regulatory authorities have not yet fully appreciated the compensatory and trust-promoting aspects of tolerance. Both in official documents (e.g. HSE 1992a) and in subsequent evidence to other enquiries, these bodies remain ambiguous about the safety net of compensation in the face of a serious accident. They also fail adequately to justify the benefits against alternative means of providing the same service. The basic conditions of risk tolerance are simply not being met.

- *The psychometric paradigm of risk tolerance*, based either on simplistic awareness–response inventories or on quantified reactions to levels of safety and regulation, is not in itself sufficient to characterize the social meaning of risk. That paradigm needs to be set in a cultural frame looking at the underlying motives and aspirations of the respondent towards the social control of technological regulation and technological advance. The marriage of the psychometric and cultural paradigms has yet to be achieved, although important new work is now being contemplated. For example, Slovic is seeking to elicit underlying "world-views" that impinge on risk aversion.

- *The social and spatial distribution of environmental danger* reflects the geography of poverty, ignorance and powerlessness. Hazardous waste facilities appear more readily to be located in poor districts occupied by racial minorities, just as toxic waste dumping in impoverished nations attracts much criticism but inadequate international safeguards. For example, Bullard (1994: 13) cites a US Greenpeace report that claims to show that communities with incinerators are 60 per cent more likely to have racial minorities and suffer property values 35 per cent below the national average. This fuels a feeling that any future risk will exhibit the evolving world of power and effective participation. Hence, the call for civic science in risk management.

- *The structure of regulatory institutions shapes risk management* more than the methods or powers available. "Institution" in this sense combines both process and organization, so the notion incorporates the interconnections between the actual design of a regulatory agency and its relationship to government, with the procedures it follows. The Royal Society report summarized seven "doctrinal contests" to portray different ways of conceptualizing regulation (1992: 159). In essence the contested positions caricatured the "old" science and a radical version of "civic" science in the management of risk. There is a middle ground, admittedly very messy and by no means an amalgam of the caricatures presented in that report. This middle ground combines structure and regulation with participatory modes and deliverable safeguards.

The common approach adopted in the US is that of mediation, now officially supported by both US industry and by the Environmental Protection Agency. Groups can hire individuals to articulate their concerns and negotiate on their behalf. This practice is beginning to take hold in Europe, with the rise of community panels connected to risky activities such as chemical plants and nuclear facilities. Petts (1994) reviews experience with community involvement as well as more participatory review panels in her analysis of waste treatment siting and management. She calls for more well monitored appraisals of consensus-seeking approaches to discover how well they score on generating trust and credibility in neighbouring populations.

To date, however, the social dynamics of consensus-seeking processes are still at an embryonic stage. The groups are consultative, relatively inexpert and effectively powerless in the tolerability sense. In an admirable review of the failure of regulatory consultative techniques to assuage local concerns over hazardous waste factors, Petts (1994) argues that reliable monitoring, open publication and rapid response to local concerns are not enough. The waste industry has to show that it is limiting its own future by promoting the cause of waste minimization at source.

Civic science may not catch on as a name. Science is genuinely fearful of losing its image of authority, integrity, tradition and public service. To swallow it up in argumentative political processes is regarded as a serious matter by numerous scientists in many disciplines. It is therefore wise, for the moment at least, to conceive of civic science as a subset of risk management, global environmental change and the sustainability transition, the three realms where its peculiarities and novelty are most required. This point is elaborated upon in O'Riordan & Rayner (1992). For the purposes of this essay, however, only the role of civic science in risk management will be given any serious attention.

The incorporation of civic science in risk management

Let us examine how far risk management as a mix of social and natural science perspectives, has evolved since the "risk revolution" began in the early 1980s. The creation of the journal *Risk Analysis* heralded an era of much greater interdisciplinarity in risk research, together with a host of speculative sociological critiques of risk regulation, all of which are admirably summarized in Chapters 5 and 6 of the Royal Society report (1992).

The notion of risk as a *combination* of technical appraisal and social tolerance is all but established, especially in the tricky areas of nuclear technology, toxic biotechnology and genetically modified organisms. How the combination is actually *amalgamated*, rather than *aggregated*, is a matter for regulatory structure and procedural innovation. Amalgamation requires outreach and a genuine commitment to negotiation. Aggregation is simply the summation of views conducted in the secrecy of bureaucratic vaults (see Pilisuk et al. 1987 for one perspective on this point; for a more general argument, see Krimsky & Golding 1992).

Advisory bodies have begun to open up their membership to social scientists, consumer group representatives and to non-governmental organization activists. This is not the politics of incorporation; it is the politics of informed debate. In the UK, all governmental advisory bodies must have an "environmental scientist" to reflect the wider understanding of civic science. Prominent UK bodies such as the Radioactive Waste Management Advisory Committee, the Advisory Committee on Releases to the Environment of Genetically Modified Organisms, and the recent publications on the regulatory style of risk tolerance published by the HSE in 1990 and 1993, all attest to the genuine desire to incorporate civic science into risk management. Levidow & Tait (1993) provide a useful perspective here from the vantage points of biotechnology. By and large, this is an area of relatively little public protest, and much credit is due to these reforms.

The "market testing" of all regulatory bodies in the UK has opened up avenues for greater internal and external accountability. This is connected to the slowly evolving significance of the "citizens' charter" as a mechanism to ensure more openness and responsiveness in public life. There is an enormous opportunity to be grasped, but the muddle created by the management changes has still to be sorted out. In principle, however, the scope for extending civic science into the operational effectiveness of all regulatory bodies is very great indeed. This would require, for example, better justification of risk–benefit analysis (which could work to the detriment of safeguards), better explanation of the evidence on which hazardousness is evaluated, and more open procedures for explaining safety levels with a workforce and the public. Such innovations will have to be fought for: ironically this is happening at a time when the government is trying to reduce regulatory strictures, when open government is still being officially resisted, and when pressure groups are growing short of cash and membership and so may have difficulty in seizing the opportunities available. This is a very new area of enquiry with, as yet, no substantial research in progress.

This essay is not designed to produce a "wish list" for research. But the creation of non-departmental public bodies run on corporate lines does deserve thoughtful scrutiny by those interested in the application of civic science to risk management. Currently (1996) in the UK, the mainline regulatory agencies, namely the HSE and Her Majesty's Inspectorate of Pollution, are being restructured along corporatist management lines. The former will rely increasingly on income from licensing procedures, funded at regional office level, with an arm's-length relationship to policy. The latter is to be incorporated into a new Environment Agency, amalgamating waste regulation and water management into a single body.

Carter & Lowe (1994) survey the merits of quasi-independent regulatory agencies and their relationship to ministers and to parliament. There is little doubt that these relationships are still in transition, with great uncertainty regarding the precise combination of political control and regulatory freedom. But again the emphasis will be on budget-holding and regulatory operation at the local level, funded from fees from issuing licences to discharge pollutants.

It is by no means clear that the kind of reforms recently developed for biotechnology and for genetically released organisms will extend to these new regimes. Ideally, some form of community advisory panels should be actively integrated into these local offices, encouraging dialogue, openness and trust at the very point of regulating operation. Curiously, this move would combine the citizen's charter and civic science into one new institutional arrangement. As Petts (1994) notes, this is an area where honest experimentation needs to be independently evaluated. But there is precious little evidence that the management consultants are thinking this way.

Civic science and emerging trends in risk management

Civic science creates a paradox. On the one hand it is designed to reduce danger in our lives, by acting out the ambiguous principle of precaution. On the other, it thrives on adaptive learning arising out of errors of judgement or unanticipated disasters. The paradox is not easily explained away. Suffice it to say that precaution may prevent some horrors, but it will never eliminate the unexpected. And genuinely unforecastable outcomes are arguably vital to "keep us on our mettle" and enhance social cohesion and community solidarity. The fact that chemicals will still deplete the stratospheric ozone layer throughout the next millennium, despite well publicized scientific prognoses and international protocols to remove or phase out some of the offending substances, does not mean either that precaution will eventually prevail or that unanticipated scientific findings and health hazards are yet to materialize. The issue of stratospheric ozone depletion carries its inherent dynamic: risk and adaptation can be made mutually supportive. One mechanism for doing this is civic science.

Comparative risk management

An area of emerging interest in civic science is *comparative risk* (Minard 1993). This is a lively activity in the US. In essence, round-tables of representative citizens rank the various environmental problems in their surroundings, and debate how best to overcome them, including common solutions. In simple terms this process means identifying, ranking, prioritizing, analyzing, implementing and evaluating. The advantage lies in facilitating a serious review of all environmental changes, placing risk within that review, and selecting solutions that drive at the heart of environmental change generally and not just risk management in particular. Obviously, this tactic requires well staffed and informed round-tables, plenty of time, competent and adaptable officials, and a commitment to follow up agreed action. These are early days, but it could be the embryo of local Agenda 21 discussions concerning the transition to sustainable development at the local government level.

Minard (1993: 18–19) reports on examples of citizens groups ranking issues in broad agreement with experts, thereby displaying common knowledge, yet putting different values on levels of investment for improvement. This has given political impetus to regulatory and management agencies to address various local issues (e.g. protection of threatened ecosystems, citizens' surveys of health problems, geographical distribution of hazard in relation to ethnic composition) with a much higher degree of political support.

Social learning and institutional adaptation
following high-profile disaster

Arguably, the 1980s were a period of challenge in risk management research. The onslaught was led by the social psychologists, who sought patterns in the social responses to danger. Because much of this work was geared to generalization and production, to assist beleaguered politicians and technical specialists who could not secure public acceptance of hazardous processes and facilities, there was inevitably a countercharge by the anthropologists and cultural theorists (see especially Rayner 1992). They argued for a much wider frame of reference, looking at how people developed structures to associate blame, to assimilate uncertainty, and to build trust in an innovative and rapidly changing technical world. The repeated shockwaves caused by mega-disasters helped all this – Seveso, Bophal, Chernobyl, *Exxon Valdez*, *Challenger*, and Sandoz.

Mega-disasters are probably necessary to advance the cause of civic science, even though, tragically, they leave many martyrs in their wake. Social learning is triggered by episodic convulsions. These highlight the inadequacies of regulatory procedures and safety standards; they throw doubt on the competence of expertise; they suggest that science has to be extended to trusted communication if it is to remain credible; and they symbolize how far a presumed benefit can be made tolerable through safety measures, technical fixes, organizational responsiveness and mobilized pressure groups.

Kai Lee (1993: 156–64) cites the American political scientist, John Kingdon (1984) on the topic of how policies are formed. Kingdon claims that policy ideas swirl around communities of analysts, advisors and politicians, coopted interest-groups and think-tanks. They become activated in a time of crisis or sudden change of perspective. Given the right conditions, policy ideas may suddenly have legitimacy and constituency, so they are put on the agenda. Mega-disasters create such a climate of serendipitous acceptance. What we can never know is how far the learning experience in the convulsive aftermath of such events creates institutional change of sufficient quality as to stop the many similar possible events that might otherwise have occurred in the future had the accident or incident not generated the learned adaptation.

The incorporation of precaution into risk management

Precaution is coming of age; arguably, it is the most exciting development in contemporary environmental science and politics (O'Riordan & Cameron 1994). In a nutshell, precaution requires action ahead of scientific proof, when cost effectiveness of response cannot be guaranteed, and when costs to the public, in terms of behaviour change, new taxation or policy shift, could cause hardship and resentment. Precaution is essentially about extending science to the public realm, about re-ordering victim powerlessness in favour of new mechanisms of victim avoidance, and of guaranteeing buffers of protected "ecological space" or "safety" to avoid going too close to unknowable but lurking thresholds. Because precaution places the burden of proof on the risk creator to show no unreasonable harm and to build in guarantees of compensation in case of honest misjudgment, so precaution favours the would-be victims rather than the beneficiaries of risk-related decisions. In the world of global environmental change, the full influence of precaution would be politically startling.

In terms of risk management research, precaution places new responsibilities on science to integrate across the natural/social science divide. This is because there are genuine uncertainties in both physical and social systems. Uncertainties of process, of laws, of data collection, of historical trend interpretation, and of responsiveness to new findings or new input relationships. Precaution presents a case for civic science type structure of interdisciplinary science, upon government, more complete communication, and dialogues between social judgement of possible outcomes and cost-effectiveness calculations of various courses of action.

This would also permit greater use of mediation techniques and mechanisms in European risk management. To date the role of mediators has been played down in Europe in favour of more open and participatory regulatory arrangements in public inquiries or hearings.

Precaution is at its most potent in areas where "gut feelings" of alarm are not assuaged by emollient scientific judgement. Examples include BSE ("mad cow" disease) and adequate safeguards over the diet of cattle and the failsafe slaughter of possibly infected animals. Yet there is a rumbling rumour that BSE may be neurologically connected to organophosphates in cattle feed, for many insecticides rely on organophosphate substances. This is genuinely unsubstantiated. But it is a sign of the modern risk debate that such connections are suggested and, although by no means proven, become legitimized in the minds of doubters. This is partly because, as noted at the outset, the apparatus of scientific proof is suspect. There is much discussion of a possible link between organochlorides and infertility, caused by the possible mimicking of reproductively inhibiting hormones in some of the organochlorine compounds. Again, science as conventionally defined cannot unequivocally provide an answer. Yet the critique is strong enough for the US Environmental Protection Agency to consider a full-scale environmental review of all chlorinated substances, and the US–Canadian Joint International Commission on the Great Lakes has urged a ban on all organochlorinated products and emissions.

In the light of this, the US Chlorine Coordinating Committee and some international chemical companies, spearheaded by Dow Chemicals, are considering banning the commercial production of substances "deemed likely to reasonable opinion" to be bioaccumulative, persistent and toxic. It will be a remarkable day when a major corporation acts ahead of regulation, in the name of civic science, to withhold some of its products. But Dow Chemicals may well be in this position within a year.

Emerging institutional structures in European risk management

As the European Community becomes more integrated, so its risk management institutions may begin to converge in terms of both procedure and culture. So far, the evidence is equivocal, because member-states are jealous of their own styles of operation. But as borders open and risk continues to be trans-frontier in character, so there may well be moves to coordinate and collaborate on risk management. Into this pot of reconsideration might be thrown many of the themes raised in this section – civic science, mediation, precaution, deregulation, cost-effectiveness and internationally agreed adaptation to mega-disaster. If all this takes place in a manner capable of being examined and commented upon, it should provide useful suggestions as to how risk management and civic science might evolve in an enlarging Europe over the next decade. Both the amalgamation and enlargement processes are taking place within a broad notion of "ecological modernization" (see Weale 1992). So, it would be timely to evaluate moves to incorporate active and informed opinion to create a wider civic partnership in risk management. To this end, case studies of the most promising experience need to be compared and evaluated. The tricky combination for the European Union is to retain a homogeneity of safety measures and risk management principles while preserving the heterogeneity of regional responses. This should be possible, but it is a combination that will require conscious promotion and an enormous amount of local support.

Conclusion

Civic science remains a theme for academic analysis and irregular attention arising from protest and frustration by NIMBY groups. Risk management cannot progress without it, however, for "risk" is inherently a civic science concept. Its progress will most likely evolve incrementally through stealth rather than by any "big bang" explosive growth. Civic science will emerge within risk management as

- advisory bodies become more vernacular and not purely scientific
- regulatory agencies adopt community advisory panels at area office level
- community groups work with local educational establishments to produce "maps" of danger, deprivation, disease and despair, so that the social justice aspects of all this are formally taken into account
- long-range siting plans take these maps into account and explicitly incorporate risk-avoidance measures
- politicians heed public alarm, expressed through legitimate channels
- economic evaluation connects value judgement to proximate monetary values and applies a precautionary spin to cost–benefit analysis
- local groups are enabled to visualize minimum-risk futures and the consequences for employment, job security, industrial competitiveness and political democracy of their proposals and yearnings
- all this is done on an integrated community level.

CHAPTER EIGHT

The regulatory target: products and structures – or people and organizations

SHOULD REGULATION BE TARGETED ON PHYSICAL PRODUCTS OR INSTITUTIONAL PROCESSES?

The final issue in the risk management debate to be explored in this book concerns whether risk is better coped with by changing physical structures or changing the behaviour of individuals or organizations. Where should the emphasis be laid between the specification of physical products or structures on the one hand and the specification of institutional processes on the other? The difference of emphasis is, in part, related to the debate about the appropriate degree of "anticipationism" in risk management which was discussed in Chapter 2. Like the institutional design issue (see Ch. 5), this dimension of risk management is not characterized by sharp and explicit debate. But there is a spectrum of possible positions which vary according to the emphasis placed on the specification of *outcomes* (in the sense of laying down physical standards) as against the specification of decision *processes*. In the traditional regulatory paradigm of natural science, risk management is conceived of as essentially about the design of products or physical structures that are safe within a specified set of functions, or at least "safe enough" within cost–benefit constraints. The emphasis is on incorporating the expertise of natural science, medicine or engineering into authoritative research-grounded standards and specifications (as with the specification of rules about maximum daily intakes of certain food additives or chemical residues in food or pharmaceutical products, or permitted levels of exposure to radiation for workers in radiology departments). Such an approach is deeply embedded in much of the institutional structure of risk regulation and its surrounding decision advice procedures.

In practice, such standards often rely heavily on socially negotiated notions of "feasibility", "practicality" and "reasonableness"; for example, in the idea that particular hazards or contaminants should be rendered "as low as is reasonably possible" (ALARP) or "as low as is reasonably attainable" (ALARA). A related goal, which balances the aspirations of science with both the practical demands of technologists and the social concerns of other groups, is that set by the principle, already mentioned, of seeking the "best available technology not entailing excessive cost" (BATNEEC). As we have noted earlier, such goal definitions for product specification hinge on it being possible to find legally or culturally agreed interpretations of what exactly constitute "reasonableness", "excessive cost", "best available" and "practicable". The known existence of cultural variability in attitudes to risk suggests that it is dangerous to assume homogeneity in such matters.

At the other end of the spectrum are those who argue that a regulatory emphasis on the design or composition of products and structures is of limited usefulness, and may at some point become self-defeating, given irreducibly high levels of uncertainty, limited opportunities for laboratory testing in some crucial areas of risk management, difficulties of extrapolation from a few data-points or animal experiments, and indeed declining popular faith in the authority of natural science (Wynne 1992c). The claim is that, in circumstances where uncertainty cannot be entirely eliminated, the traditional physical-standards approach needs to be supplemented or even replaced by an emphasis on specifying organizational *processes* that will ensure the careful balancing of arguments. Such an approach to organizational design is exemplified by the philosophy supporting BS5750 (British Standard 5750), which claims to offer a standard for quality assurance within corporations. Those who favour the process-based approach argue that the emphasis must inevitably be placed on structuring the way that decisions are taken, rather than on specifying physical output in circumstances of inherent uncertainty – as in those "trans-scientific" issues referred to in Chapter 7, which are in principle resolvable through systematic investigation of hard data but where such experiments cannot in fact be done because of time, resource or legal/moral limits.

It is on the basis of such arguments that Majone (1989), who developed his analysis from an initial research interest in Bayesian statistics, put the case for a process-based regulatory approach through devising forums in which competing claims can be advanced in a manner that deters the "capture" of public policy by a particular scientific school without effective challenge and to allow well informed decisions to be taken in circumstances where scientific certainty cannot be attained. However, the development of an organization–process approach to risk management is not necessarily incompatible with a structural design approach, and the former is often laid on top of the latter.

This debate is neatly examined in the following section by Simon Shohet with references to the regulation of biotechnology in Europe. After a wide-ranging discussion of the issues, he concludes that process-based precautionary legislation is the preferable option in this particular case.

RISK AND EMERGING TECHNOLOGY: THE CASE OF PROCESS-BASED REGULATION OF BIOTECHNOLOGY IN EUROPE

Simon Shohet[1]

Introduction

The advent of biotechnology and the techniques of genetic modification (GM) have created a range of policy dilemmas for governments. Tensions have resulted from the need to safeguard human health and the environment from perceived hazards while ensuring that regulatory frameworks do not constrain innovation and wealth creation (see for example, Shackley 1989, Levidow & Tait 1992, Tait & Levidow 1992).

The precautionary principle has dominated European regulation of GM – hence the process or technology (and its use in research, development and manufacture) is the focus for initial regulatory supervision, rather than specific products. However, the validity of various regulatory options has been a matter of fierce debate and heavy lobbying, both by industry and environmental groups. This paper examines the arguments and reviews some recent developments, drawing on examples from the UK, which introduced voluntary codes of practice for genetic manipulation in the mid-1970s. Subsequently, it took a lead in implementing two European Directives in the early 1990s. The directives set a legal framework for contained use, environmental release, and marketing of genetically modified organisms (GMOs) and derivative products.

A key trigger was the report of The Royal Commission on Environmental Pollution (RCEP 1989). It came down in favour of specific regulation of biotechnology, the essence of which was accepted by the UK government and led to enabling legislation under the Part IV of the UK Environmental Protection Act (1990). However, in 1993 the House of Lords Select Committee on Science and Technology initiated a review of the impact of regulation of biotechnology on the competitiveness of UK firms (House of Lords 1993). This was strongly critical of the RCEP and the European Commission, and argued that the rules were excessively cautious and bureaucratic. The basis of these opposing views is examined critically in this paper, and a case is presented to support the RCEP's recommendations and the initially cautious approach of the European Commission.

---

1. The author is grateful to the Gatsby Charitable Trust for research support. The views expressed in this article remain the author's and may not necessarily reflect those of the sponsoring organization.

Background

Genetic manipulation in this context can be described as the use of recombinant DNA techniques to alter the existing genetic complement of an organism – to insert, remove or rewrite one or several of the genes that make up the naturally occurring or domestic version of the organism concerned. Genes comprise DNA, which holds the assembly information for proteins (including enzymes) that determine the structural and functional components of organisms. Some unique features of GM, outlined below, must be considered in any evaluation of policy options for risk assessment and regulation.

Of major importance in the context of the release of organisms into the environment, whether accidental or deliberate, is that natural selection pressures will immediately begin to operate. Uniquely, risk assessment in such situations will require not only consideration of the organisms as individual entities but also of the information (DNA) that they have inherited and will pass on. That information enables an organism to make copies of itself and potentially, at least, to colonize new environments. Of course, it is precisely this property of living things that is exploited in agricultural practice. But the unique feature of living organisms, as researchers have pointed out (Fincham & Ravetz 1990, Munson 1995), is the capability for magnification as populations of organisms multiply, colonize and adapt over time – a factor absent from purely chemical and physical environmental disturbances. Thus, the old adage that "the solution to pollution is dilution" does not hold for self-replicating organisms.

When the techniques for generating recombinant DNA (rDNA) first appeared in the mid-1970s, public concerns – and indeed the initial concerns of scientists themselves – resulted in the establishment of national supervisory committees such as the Recombinant DNA Advisory Committee (RAC) of the US National Institute of Health and the Genetic Manipulation Advisory Group (now the Advisory Committee on Genetic Manipulation, ACGM) in the UK. Since then, however, knowledge in key disciplines – such as genetics and molecular biology – has advanced enough that those wishing to commercialize the products have been able to argue that it is timely to move from cautious supervision of the process towards assessment of the products irrespective of the technology used in their production. Indeed, release and product approval in the absence of specific laws covering genetic manipulation has been the approach of the US Federal Administration and the administration in Japan (House of Lords 1993: appendix 4). In contrast, the European Commission, which under Single Market regulations has executive powers over all 15 members of the European Community, has preferred the route of process-based regulation specifically aimed at the technology, rather than its outcomes.

Contentions in the risk-regulation debate

Views are polarized (Table 8.1): industry and many scientists have argued that biotechnology and genetic manipulation techniques will bring major economic and social benefits, such as new therapeutic drugs and improved crop varieties. They contend that there are no scientific reasons why genetic modification should pose additional risks to human health or the environment.[2] As such, regulation should be flexible and supportive of technological innovation and development in this area. Those opposed to the techniques, including some environmental groups, argue that, by crossing species boundaries, genetic manipulation is tampering with nature in a way that traditional selection techniques have not done.

Table 8.1 Summary of technical contentions in the release of genetically engineered organisms into the environment (see also RCEP 1989; Wheal & MacNally 1990, Munson 1995).

Pro-release	Anti-release
Genes are well understood entities.	Gene actions are poorly understood. Microbial interactions and systems are highly complex.
DNA movement outside species is limited.	There is evidence for genes crossing the species boundary in both microbes and crops.
Risk assessment procedures adequately identify human health and environmental safety hazards.	Scientists have adopted a narrow participationist approach, i.e. "we know, you don't". Risk assessment is subjective based on previous experience, which is largely incomplete.
Selective breeding of crops post-GM eliminates undesirable traits.	Some crops, such as oil seed rape, have certain weed-like characteristics.
The technique of GM is a highly controlled form of natural selection.	GM is fundamentally different from traditional techniques.
Products still have to be assessed on the bases of quality, safety and efficacy, which provide a fail-safe filter for rogue products.	The technology at research, development and manufacturing levels needs regulatory supervision, not just the product outcomes.
There is no evidence to date of damage to the environment or human health caused by GMOs.	Environmental and health liability laws in the UK are relatively weak. Use of genes conferring herbicide resistance in crops will lead to increased use of chemicals.
In vivo monitoring techniques are available.	Monitoring techniques are inadequate.
Regulation of the technology is "unfair discrimination". For an industry to be competitive (e.g. in the UK) regulation must be no more stringent than in rival countries (e.g. in the US).	Regulation that is seen by the public to meet its concerns reassures the public as consumers; this is good for the industry. Concerns about the "level playing field" have not been substantiated with evidence.
An average higher organism has between 50K and 100K genes. A single insertion means that over 99.99% of the genome is unchanged.	A single gene can theoretically confer physiological properties such as toxicity and pathogenicity.

2. See, for example, House of Lords (1993: 73).

First, these actions will raise a range of environmental and health risks and, secondly, they generate certain ethical considerations (Haerlin 1990). The latter point is clearly important and arguably has influenced the framework in which European regulation has been set; however, it is beyond the scope of this section (RCEP 1989: chs 4, 5; Wheale & MacNally 1990, Mepham 1994, Munson 1995).

Product regulation

In very general terms, in the absence of knowledge about the risks of an emerging technology, there is no *a priori* case why process regulation should be favoured over product regulation or *vice versa*. To illustrate this it is clear that, in some cases, hazardous processes can produce relatively low-risk products (e.g. mining or quarrying), whereas in other cases, relatively simple processes can produce dangerous or toxic products, for example, the extraction of narcotics or toxins from plants, or even the breeding of dangerous dogs.

From a business perspective in an R&D intensive industry with high sunk costs, product and sectorally targeted regulation offers firms some certainty that, provided criteria are met, a fair regulator will judge products on their own merits. In the case of bio-pharmaceuticals, for example, these are the established criteria of quality safety and efficacy. Safety concerns are theoretically exposed in tests and trials of potential products. Additionally, many practitioners have argued that process-based regulation discriminates against biotechnology, even though the end products from biotechnology-derived routes and traditional routes may be very similar or identical. For example, certain amino acids can be manufactured by either hydrolysis of proteins or by fermentation using genetically modified microbes such as *Corynebacterium glutamicum*. However, where process-based regulation operates, the products would require separate regulatory assessments, despite being chemically identical.

After a careful analysis of safety considerations for rDNA organisms, the OECD (1986: 41) concluded that

> there is no scientific basis for specific legislation for the implementation of rDNA techniques and applications. Member countries should examine their existing supervision and review mechanisms to ensure that adequate review and control may be applied while avoiding any undue burdens that may hamper technological developments in this field.

The US administration arrived at similar conclusions in 1991 (US President's Council on Competitiveness 1991). In other words, the process of GM in itself does not warrant special regulatory supervision.

Background to current process regulation of biotechnology in Europe

Those institutions favouring process regulation have argued that in the case of biotechnology there are inherent risks in the genetic manipulation process, which necessitate that regulation begins at the act of genetic manipulation. It is argued that the role of regulation in the early stages of the process does not inhibit innovation or restrict R&D practices as practitioners have suggested, but rather that it stimulates and drives innovation; thus, firms that are well placed to take the lead in meeting higher technical or environmental standards have market advantage (for a presentation of this argument see, for example, Fisk 1993). In the case of a powerful technology where the basic equipment and reagents are easily available, it can also be argued that tight regulation "keeps out the cowboys", which benefits the rest of the industry. There is the additional case, as noted above, that regulation can mollify public distrust.

Despite strong lobbying from many applied scientists and multinational firms represented through the Senior Advisory Group on Biotechnology (SAGB 1989), the European Commission proceeded in 1990 with process-based legislation that required all genetic manipulation to be notified and in some cases prior authorization to be sought from national authorities. All environmental releases required prior authorization with waiting periods of up to 90 days (EC Directives 219/90 and 220/90) before a consent could be granted.

The highly influential report (RCEP 1989) of the UK Royal Commission on Environmental Pollution recommended strict regulatory controls for the use of genetic manipulation, many of which were incorporated into UK national law under part VI of the Environmental Protection Act 1990 and subsequently into detailed secondary legislation under the Genetically Modified Organisms Regulations 1992 covering contained use and deliberate release. These latter regulations also implemented EC Directives 219/90 and 220/90 into UK national law. As part of this implementation process the UK Advisory Committee on Releases into the Environment (ACRE), was made a statutory body. This body consists of industry and employee representatives and academics, with a secretariat from the Department of the Environment as well as observers from other relevant government departments. Its purpose is to oversee releases on a case-by-case basis and to ensure that risk assessments are evaluated appropriately on scientific grounds, but, as Levidow & Tait (1993) point out, it also has an implicit public interest role.

Other EC member-states have a legal obligation to implement EC Directives 219 and 220 and, at the time of writing, have prepared legislation or fully implemented these rules within their national laws. In other countries outside the European Union, the picture is rather different: Japan operates a voluntary system adhering to the detailed recommendations of the OECD of 1986 (Munson 1995), and in the USA federal and state legislation exists, although there is no specific "gene law" (House of Lords 1993: appendix 4).

The debate on product versus process legislation

In nearly all countries where the issue has arisen, private firms, have on the whole, lobbied strongly for sector-based legislation (sometimes referred to as vertical legislation) that can be applied to the particular uses of the technology and which will separate, for example, genetic modification in the pharmaceutical sector from that intended for agriculture. Conversely, they have largely been opposed, as already discussed, to so-called horizontal or generic regulation (Table 8.2; detailed statements of views can be found in House of Lords 1993).

Table 8.2   Some broad differences between so-called vertical and horizontal regulation.

Vertical	Horizontal
Narrow technical definitions used in setting legislation	Broad technical definitions used in setting legislation
Sector or industry specific: narrow boundaries of applicability	Legislation is cross-sectoral with wide boundaries
Aimed at products	Aimed at the technology or process
Civil administration of laws likely to be devolved down to relevant sector ministries	Civil administration likely to be centrally coordinated
Risk assessment relates to intended use.	Prior assumption of same risk irrespective of use

It has been argued that "product-based regulation is to be preferred whenever practicable on the grounds that it is better targeted and more economical for both regulators and the regulated and it does not single out genetic modification unnecessarily for a different style of regulatory treatment" (see House of Lords 1993: 58).

Against this, regulators can point to the benefit to industry of a single "post box" for handling regulation that is most efficient when administered centrally via "horizontal" legislation.

The polarization of views is well illustrated in the conflicting recommendations of the Royal Commission on Environmental Pollution, (1989) and the House of Lords Science and Technology Committee (1993). Some of the key points of disagreement that relate to fundamental scientific assumptions and risk perception are outlined in Table 8.3.

Table 8.3 Illustration of points of disagreement between RCEP (1989) and House of Lords (1993).

House of Lords (1993)	Royal Commission on Environmental Pollution (1989)
"Hazards identified by environmentalists, by the Department of the Environment as regulatory authority and by the RCEP are conjectural. Concerns are based on false analogies . . . except for a few exceptions deliberate release of GMOs is not inherently dangerous."[a]	"A major cause for concern and of expense in the release of genetically engineered organisms is uncertainty as to the effect . . . on organisms, ecosystems and the environment at large"[b] . . . "Some organisms . . . in the extreme . . . could have serious environmental consequences."[c]
"Except for pathogens, separate regulations in contained use is unnecessary."[d]	"There will need to be an extension of controls over contained work on genetically engineered organisms to minimize the risk of damage to the environment."[e]
"In framing the directives on which the UK regulations are based the European Commission took an excessively precautionary line, which in terms of scientific knowledge was already obsolete when they were being prepared in the late 1980s."[f]	"The European Commission has published draft . . . proposals for regulation of experimental releases, which have many similarities with our own."[g]

a.(House of Lords 1993: 54, para. 6.10).
b.RCEP (1989: 48).
c.RCEP (1989: 84).
d.(House of Lords 1993: 54, para. 6.9).
e.RCEP (1989: 96).
f.(House of Lords 1993: 55).
g.RCEP (1989: 69).

Which regulatory policy is the right one?

The task for policy-makers has not been an easy one; how do you regulate a powerful generic technology without stifling innovation and the development of useful products? As has been shown, expert opinion on this has differed widely from those that argue that, as the House of Lords put it, "short of an act of evil genius . . . genetic modification should not be singled out from other experimental work" (House of Lords 1993: 53), to broad concerns – as expressed by the Royal Commission on Environmental Pollution (1989) – that "organisms which survive and become established could affect the environment in several ways – both beneficial and undesirable. Some organisms could pose a threat to human health and conceivably affect major environmental processes." (RCEP 1989: 18).

Recently, political pressure and heavy lobbying appear to be shifting the European policy agenda towards product-based legislation that could enable products to be commercialized more freely, while loosening regulatory supervision on research, development and manufacture. In the UK, interdepartmental battles have been fought between the Department of Trade and Industry, which has been charged by the Prime Minister with removing regulatory burdens from industry, and the Department of the Environment, which has championed the precautionary principle and horizontal regulation of biotechnology.

Drawing all of these features together, it is clear that the issues turn on the implicit burden of proof each faction is willing to accept, in other words, whether risk regulation should be based on anticipation, with emphasis on detection and prevention, or towards resilience, where the emphasis is on the capacity to cope with the unexpected via rapid action (Hood et al. 1992). Tait & Levidow (1992) point to the complexity and uncertainty in GMOs and their interaction with the environment, which leads to unpredictability. This is the proactive rather than reactive approach built on the German *Versogensprinzip* (or precautionary principle), which has had a significant impact on northern Europe, but as Hood et al. (1992) argue (see Ch. 2), the precautionary principle does not take account of economic cost–benefit considerations. Strict financial penalties and liability can be used as a disincentive for risk-taking, but in the case of biotechnology the UK Environmental Protection Act 1990, although placing a duty of care on companies, lacks a well developed case law. The disincentive is therefore signalled only weakly. Companies have lobbied hard against heavy environmental liabilities in this area, despite having complete confidence in their products. Munson (1995) identifies the complexities of establishing liability requirements for GMOs and the general lack of consideration of liability issues in legislation when it is framed. Ideally, lessons about previous failures and an element of "forethought" about future ones can be used to make judgements about the right levels for liability, but in the case of releases of GMOs there is only the fairly limited history of experimental deliberate release to draw on, some of which has been inconclusive. The introduction of alien species was considered by RCEP (1989) to be a valid analogy to the release of GMOs and was used, among other things, to justify their acceptance of the case for precautionary legislation in the UK. The UK House of Lords strongly rejected this analogy (House of Lords 1993).

The House of Lords review placed little importance on the context dependency of technological risk. Although genes can now be well characterized, the complex interactions of the environment in which they are placed remain poorly understood. Examples abound where well "codified" entities have interacted with a complex uncodified environment in unpredictable ways, with occasionally catastrophic results. For genetically modified micro-organisms, this is particularly crucial, because recovery and remediation may be extremely difficult.

In fact the OECD report on recombinant DNA safety considerations made these same points (1986: 25):

A key difficulty is the assessment of interactions of the micro-organism with the existing ecosystem. For example, an introduced micro-organism could transfer genetic material to other micro-organisms. Once established, micro-organisms can potentially alter the environment in ways that promote further proliferation of genetic transfer, giving rise to secondary effects.

The report went on to discuss the possibility that recombinant organisms may possess unique characteristics which set them aside from other variants:

Using recombinant DNA techniques, very specific modifications can now be introduced into organisms, and barriers that have previously restricted the transfer of genetic material between species can be overcome or circumvented. Some of the micro-organisms, plants and animals produced using rDNA techniques may, thus, differ qualitatively or quantitatively from the variants found in nature or developed through conventional breeding activities. (ibid.: 28)

However, these concerns were not reflected fully in the final recommendations of the OECD report which came down against specific rDNA regulation.

The UK House of Lords report, being highly critical of both the UK Department of the Environment and the European Commission, has provided a powerful lobbying tool for industry. However, the House of Lords report acknowledges that its investigations were initiated "following allegations by industry" (House of Lords 1993: 9). In this sense the report lacks impartiality and, although it criticizes the RCEP (1989) for being "conjectural" (ibid.: 54) and the European Commission for using scientifically "obsolescent" information (ibid.: 55), it cannot itself be regarded as a balanced techno-economic analysis, since no independent economic research was carried out and instead it relied heavily on practitioners' perceptions. There is little evidence that EU regulatory policies have affected the R&D investment decision of non-European firms and, although incumbent firms have often threatened to withdraw their research activities because of restrictive legislation, with the exception of those exposed to strong public resistance in Germany, few are likely to withdraw major investments on these grounds alone.

Ironically, although the House of Lords report may have the desired effect of influencing European regulators towards product-based legislation, there is the real possibility that it has also served to magnify perceived fears of excessive regulation in the UK and the rest of Europe, thus sending precisely the wrong kind of signal to potential investors from across the Atlantic.

Conclusion

It is accepted that most rDNA products will be of low or negligible risk and therefore legislation must be flexible enough not to constrain diffusion of the technology. This can be achieved by a precautionary structure that permits reclassification of organisms in the light of scientific progress on a case-by-case basis, and by ensuring that the institutional supervisory structures are efficient, thus avoiding product approval delays and excessive costs. This can be thought of as "guilty until innocent with parole for good behaviour". At least in principle, the European Directives 90/219 and 90/220 were framed with this in mind, allowing clearance of products and provisions for technical amendment throughout the European Union. On these grounds alone, the process-based precautionary legislation adopted by the EU can be justified. Arguments that there should be a wholesale reversal of this approach remain unconvincing, even in the light of scientific and technological advance.

CHAPTER NINE

Conclusion:
learning from your desk lamp

HOMEOSTATIC VERSUS COLLIBRATIONIST[1] APPROACHES TO RISK MANAGEMENT

Each of the seven areas of debate reviewed in the preceding chapters raises serious issues for risk management, which recur across different specialisms and areas of policy, albeit with differences in precise terminology and emphasis. It is not claimed that the seven opposing positions are either necessarily mutually exclusive or collectively exhaustive. For example, the issue of whether to adopt a "statist" or "non-interventionist" approach to risk management underlies many of the debates in the field and might well be identified as a further separate dimension. But the seven dimensions that have been described do cover many areas of the contemporary debate and it seems likely that, in many areas of risk management, an approach that leaves any of the positions out of account is likely to be inadequate.

Many of the seven sets of risk management doctrines are in principle independent of one another. For example, it is possible to combine an emphasis on process-based regulation with either a "broad" or "narrow" position on participation in risk management decision-making. And both can be combined with either a "complementarist" or "trade-off" position. As a consequence, a large array of possible combinations of positions exist, even on these seven dimensions. But in practice some positions do tend to be readily combined with others. For example, those who favour "anticipation" are unlikely to be "agnostics" over the possibility of institutional design (although those who would place the emphasis on "resilience" may well also advocate institutional design). Similarly, those in favour of broad participation are unlikely to be pure "quantificationists". It may well be that the seven areas of debate could ultimately be reduced to some more basic set of distinctions, such as the well known quadrants of the cultural theorists (Thompson et al. 1990) as recently elaborated by Adams (1995).

Moreover, another broad thread can be distinguished as running

---

1. See p. 206 for explanation.

through many of these debates. Applied control theory distinguishes between what Dunsire (1990) terms *homeostatic* and *collibrationist* regulation processes. A "homeostatic" form of control uses feedback processes to achieve pre-set goals, whereas collibration has no agreed goal, but works by "making extremes meet" through "opposed maximizers" set up to pull a system in different directions at once, such that the system's state at any one time is a product of the interactions among the various forces, which are held in opposed tension, like the springs in a desk lamp (Dunsire 1978: 181, 207–8; 1986). Another way to visualize the two approaches is as the difference between a race from a specified starting point to an agreed finishing line, as against a tug-of-war.

In risk management, what might be termed the "homeostatic" approach places emphasis on institutional capacity to set determinate goals (a "finishing line") in advance and to convert those goals into quantified decision rules which experts can apply to particular cases, and organizations can incorporate into their standard operating procedures. Such an approach will tend to favour anticipation, quantification and the specification of outputs. It will broadly link with what Majore (1989: 12ff.) terms "decisionist" approaches to public policy as against process or institutional design approaches to shaping policy. "Decisionism", a term originally coined by Shkler (1964), means calculated choice according to a generalized decision-making logic.

The alternative "collibratory" view holds that inherent scientific uncertainties limit the possibility of reliable forecasting in many crucial areas (particularly in respect of slow-onset hazards such as global warming), and that cultural variety and dynamics limit the capacity for robust aggregate goal-setting to be elaborated into precise technocratic decision rules. The finishing line cannot be seen in advance, and hence there is no way of knowing which way to start running. The implication is that the process of managing risk requires the design of institutions (at both corporate and public management level) on the principle of the desk lamp, rather than the thermostat, by explicitly juxtaposing rival viewpoints in a constant process of dynamic tension with no pre-set equilibrium (see Schwarz & Thompson 1990). Hence, the collibrationists' position will tend to favour "resilience", specification of process and qualitative debates over uncertainties, rather than the "homeostatic" view. In a broader sense, perhaps, the "collibrationist" position might even suggest that the seven areas of debate that have been identified need not be finally resolved in one way or another, but institutionalized in the process of risk management in a way that keeps the rival positions in opposed tension – the tug-of-war.

It would probably be fair to describe the "homeostatic" view as the current orthodoxy of risk management in the scientific and practitioner community, and the "collibrationist" view as much less widely accepted. Nevertheless, it must be recognized that the orthodox position has been increasingly challenged since the mid-1980s, and "collibrationists" argue that their vision of control more accurately describes the underlying processes involved. These opposing views and a possible way forwards are discussed by Christopher Hood in the final section of the book.

WHERE EXTREMES MEET:"SPRAT" VERSUS "SHARK" IN PUBLIC RISK MANAGEMENT

Christopher Hood[2]

"I'll hae nae hauf-way hoose, but aye be whaur
Extremes meet – it's the only way I ken . . ."
(Christopher Murray Grieve [Hugh MacDiarmid], *A Drunk Man Looks at the Thistle*)

"No workable alternative" to conventional risk engineering?

As shown by the "four chapters good, two chapters bad" episode over the 1992 Royal Society document (discussed at the outset in this book), conventional "enlightened engineering" approaches to public risk management have been assailed from several quarters. But the powerful countercharge from the beleaguered champions of "enlightened engineering" goes that those who attack the orthodox model have nothing workable to put in its place. All the challengers can offer, it is claimed, is essentially negative carping criticism and a few vague and ill defined ideas, most of which are directed at how to explain behaviour and attitudes rather than how to manage complex public risk issues. Consequently, the argument goes, "enlightened engineering" orthodoxy remains the only well worked-out and publicly defensible approach, and deserves to be the central instrument of social risk management, *faute de mieux*.

This challenge is important, and it is not easy to answer. This contribution is only one of several contemporary attempts to sketch out alternatives. It aims to look at risk management through the spectacles of control theory and political science, contrasting the conventional "enlightened engineering" approach with an alternative approach built on institutionalizing rival values in risk management and keeping them in opposed tension, so that value-clashes become explicit and the balance of forces is more readily steerable by light pressures. The "opposed maximizers" principle is based on a theory of institutional control and policy intervention, which can be "viable" in both a political and control-theory sense.

---

2. I am grateful to Mary Douglas, Andrew Dunsire, Tom Horlick-Jones, David Jones and Nick Pidgeon for very helpful comments on the first draft of this chapter.

The conventional "SPRAT" approach to risk management

The conventional "enlightened engineering" model conceives public risk management as working rather like a thermostat – in what control theory terms a "closed loop" system. For convenience, that approach to risk management is here labelled as "SPRAT" (to stand for "social pre-commitment to rational acceptability thresholds"). For SPRAT, the emphasis is laid more on rational decision methods than on other dimensions of institutional design. The underlying design problem is for "society", guided by the best available scientific consensus, to somehow settle on the appropriate settings for the risk engineers to programme into the "thermostat". That is, critical levels of risk acceptability or tolerability (with all the perplexing value-of-life conundrums that risk–benefit analysis produces) need to be specified, so that levels of public risk can be kept within satisfactory bounds. To be counted as "rational", such settings need to rest on scientific and bureaucratic norms, such as toxicological margin-of-safety conventions, risk–benefit analysis or imputed risk tolerances arrived at by the Chauncey Starr (1969) method of inferring general risk acceptability by "reading across" from cases such as driving or smoking, where risky activities are undertaken by large numbers of people. It is the job of the "engineers" (through dose–response experiments, construction of fault trees or analysis of historical data) to ensure that the thermostat is capable of detecting all the conceivable sources of risk and that the mechanisms for corrective action operate smoothly.

From this viewpoint, if SPRAT breaks down, it must mean that the initial risk settings are wrongly specified, that the control system fails to sense deviations from those settings effectively, and/or that mechanisms for correcting detected violations are inadequate. When such failures occur – as they often do, given that many assessments of "risk" are in fact subject to deep systemic or parametric uncertainty (Shrader-Frechette 1991: 30) – proponents of SPRAT argue that the answer is to develop a better "thermostat". If events defy expected risk patterns, as in such cases as Three Mile Island and Chernobyl, the aim should be to make the system more sensitive, more reliable or more sophisticated, not to abandon the goal of thermostatic control altogether. After all, when computers break down or software fails, the conventional response is to remedy the system or develop the software, not to go back to slide-rules or pencil and paper.

Many of the common criticisms of orthodox risk management point to the various ways in which the SPRAT closed-loop approach tends to fail – forms of failure that could, at least in principle, be corrected by constructing a better "thermostat". But this chapter argues that there is a different way of organizing public risk management, which is not just a matter of refining the settings for the "thermostat" model. The alternative starts from a different basic style of control. It is argued here that the alternative is likely to be more appropriate to some well known circumstances (such as high politicization, no general consensus on safety goals, scientific uncertainty or trans-scientific issues) in which SPRAT tends to fail.

An alternative "SHARK" approach to risk management

An alternative way of managing risk is built on a quite different metaphor. Instead of setting a thermostat, risk management is seen as an institutionalized "tug-of-war" between incompatible pressures, with a balance-tipping mechanism. Take the case of food safety policy. The SPRAT approach is to make a definitive scientific–bureaucratic decision about what is or is not safe food through ever more refined toxicological conventions. But an alternative approach is to set up institutional decision processes in such a way that the conflicting values in play are publicly debated, for example by separating food safety responsibility from sponsorship of food production in the state bureaucracy or by subsidizing food consumer groups to challenge producers, leaving it to the political process to swing the balance.

The second approach is more like the human body's water balance system or its parasympathetic and orthosympathetic nervous impulses – operating through open-ended tug-of-war mechanisms – rather than its thermostat-style temperature control mechanism, which has (according to some physiologists, at least) a determinate "setting". Instead of determinate thermostatic settings resulting from rational decision procedures, tensions are built into institutions, with control exercised by procedural constraints rather than by output settings. The essence of "managed competition" systems is that they work by "making extremes meet" in continual struggle, with "all to play for". For convenience, the alternative risk management model is here labelled the "SHARK" approach ("selective handicapping of adversarial rationality and knowledge").

The rest of this chapter assesses the SHARK managed competition approach against the conventional SPRAT method. It aims to develop ideas about risk management through "collibration" that were floated briefly in the 1992 Royal Society document (Hood et al. 1992), but too sketchily to make the case convincingly. The claim is that SHARK is viable (in both a cybernetic and political sense) as a basis for public risk management in at least some circumstances. The institutional requirements for reconstructing public management for the stable operation of such a system are extremely demanding, but that does not apply to a more incremental strategy of redesign or to an opportunistic shifting-site approach to its operation.

Strengths and weaknesses of
SPRAT-type management systems

As noted earlier, SPRAT is an essentially "homeostatic" approach to risk management. In such a control system (according to Dunsire 1990, 1992), a pre-set datum line, defined as "acceptable risk", marks the preferred goal of management, and negative feedback mechanisms (inspectorates, hotlines, reporting and surveillance systems) are set up to compare the state of the system with the datum line and to make changes as the system starts to swing off-limits. Unlike direct steering systems (in which power is used directly to correct deviations from the desired state; Dunsire 1992: 24), homeostatic control systems separate the process of policy-setting, monitoring and active intervention to correct deviation. Like the marine self-steering devices that have replaced the traditional helmsman, they need to involve a high degree of self-regulation.

Like any other form of management, homeostasis has strengths and weaknesses. Its "engineering" advantages over direct steering are often stressed in cybernetic analysis. Those advantages stem from the inherent difficulty of achieving "requisite variety" (i.e. matching the complexity of the system to be controlled) in any control mechanism that has no element of self-regulation. Such a system will inevitably be defeated by informational variety when it operates on any extensive scale. Moreover, homeostasis clearly has some major political attractions compared with any direct steering approach to risk management. In particular, it fits with the widespread desire of elected politicians to cope with their personal political risk by avoiding direct responsibility for public risk management whenever political "credit slippage" is likely to be outweighed by the value of "blame shift" (see Fiorina 1982). The political logic – very clear in cases such as food safety – is to entrust as many controversial risk management decisions as possible to "representative" quangos or government-approved expert groups, thereby deflecting blame away from politicians onto scientific experts and committees of the "great" and the "good".

However, like any control system, SPRAT also has corresponding weaknesses. Three of its defects are familiar in risk management debates and are summarized in Table 9.1. First, and most important, SPRAT assumes a capacity on the part of "society" to agree on pre-set goals about matters of risk and blame that runs against all our knowledge of how social and political systems actually work. The unrealism of this assumption does not simply stem from the problem, classically expounded by March & Simon (1958) and their many followers, of "bounded rationality" (i.e. the cognitive limits that prevent individuals and institutions from adopting contingency plans or contracts for all possible states of the world), which limits such goal-setting capacity. It also runs up against the most basic logic of political activity.

Table 9.1 SPRAT-type risk management: elements and limits.

Control element	SPRAT model	Limits/pitfalls
	Stable agreed targets	Bounded rationality and cultural variety
Director (sets goals)	Danger thresholds operationalized in advance	Requisite variety, noise, opportunism
Detector (checks course)	Self-correction mechanisms comes into play at the point that the system begins to swing off-limits	Without multiple thresholds, danger of seizing-up propensities and implementation deficits
Effector (changes state of system)		

In politics, general agreement on basic goals is always far harder to achieve than agreement on specific measures. Groups who will never agree on basic objectives – because their goals are different or diametrically opposed – can nevertheless often agree on particular courses of action. That is how politics works, and any management system that demands agreement on goals as a prerequisite for any other activity will tend to be either cosmetic or unworkable. Take the typical case of managing the risks associated with tobacco smoking, a problem for which appropriate regulatory measures have been hotly debated for over 30 years. The basic stakeholders in this policy domain – tobacco companies and libertarians who oppose government interference in markets or individual behaviour on the one side, public health institutions and anti-smoking lobby groups on the other – will never agree on what the basic goal of public policy over smoking risks should be. But they may well give common consent or support to particular "half-way-house" measures of risk regulation, such as restrictions on tobacco advertising, precisely because such measures fit incompatible goals. Those who basically oppose government interference in tobacco smoking can support advertising restrictions in the belief that such measures will serve to stave off demands for more extensive policy, whereas those who want to see heavy sanctions against smoking can support the same measures in the belief that they will be a first step in establishing an unstoppable momentum towards more radical solutions. It is the inherent difficulties of arriving at a goal consensus in political systems that are conventionally thought to lie behind commonly occurring components of everyday political behaviour, such as "serial disjointed incrementalism", "partisan mutual adjustment" (patterns of decision-making that avoid agreement on basic objectives) and the tendency of large coalitions to dissolve over time (cf. Lindblom 1959, 1965, Riker 1962, Hood 1989).

Accordingly, any "depoliticized" management system that depends on stable goal consensus will go against the inherent grain of political life in this sense. This point has often been made in relation to "rational" (goal consensus) management systems for budgeting, land-use planning or corporate management (cf. Wildavsky 1971). Those problems apply *a fortiori* to risk management, if there is anything in the cultural theorists' argument that risk is the primary element in an (increasingly?) politicized process of blame among competing world-views (see Douglas 1992, Thompson et al. 1990). So, any approach to public risk management that starts from assuming social capacity to arrive at a stable consensus over pre-set levels of acceptable or tolerable risk, is like saying "if you can solve your problems, you can solve your problems" (see Schlesinger 1967). It states the difficulty rather than offering any solution.

Moreover, the SPRAT closed-loop model can be problematic in other ways. Even if a generally agreed level of acceptable risk could be arrived at on a stable basis, there are basic problems of "requisite variety" and distinguishing "signal" from "noise" bound up with the detection capacity of such a system – that is, the processes by which deviations from accepted risk settings are registered and monitored. After all, the disaster literature is full of astonishing stories of detector failure, cases in which obvious (with hindsight) signs of the onset of major system failure have been unaccountably ignored during a critical "incubation period" (cf. Turner 1978, Perrow 1984). During that incubation period, latent interaction pathways appear between elements of a system and escape the notice of managers or regulators, because they link subsystems that were assumed to be independent, thereby creating many more degrees of freedom in the overall system than had been expected. The social factors associated with the failure of risk management systems may be complex (see Wagenaar & Groenwold 1987, Horlick-Jones et al. 1993), but such failures often seem to be rooted in everyday traits of organization and bureaucracy, such as information asymmetries, incentives to distort information, accumulations of minor errors and slackness, communication gaps and factional feuds.

If that is true, the more the process of setting the "risk thermostat" is socially separated from implementation and detection processes by organizations, with all their real-life imperfections, the more likely it is that detector and effector failure will occur. When there is a long linkage between the processes of goal-setting and processes of implementation on the ground, and when that goal-setting takes place through a top-down process (as typically arises in public risk management, where "expertized" settings are mandated for general application), it produces the classic conditions for an "implementation deficit" of the kind so often documented in the policy-process literature (see Hood 1976) and dramatically illustrated by the case of the Chernobyl engineers who turned all the plant's safety systems off and thereby triggered the world's worst nuclear accident in 1986. The reality of risk management on the SPRAT model is that those down the line are likely to view the "official" risk settings as inappropriate or unnecessary, and that risk management desiderata will be competing with other political and economic pressures on institutions.

It may be that such difficulties are not inherent in SPRAT. In principle such a system could be set up to detect and respond to infinitely small deviations, by interposed thresholds. But where such a risk management system is built on a single dichotomous safe/unsafe threshold (as with food, drugs, responses to natural disasters and other low-probability high-consequence events), other familiar possibilities for social friction can arise. That is, if the single threshold is seldom crossed, the management "switching" mechanisms that activate the corrections will inevitably tend to "corrode". And such corrosion is not just an engineering problem of the physical seizing-up of machinery and equipment that is seldom used, although that can happen too. More importantly, it is a social and political process, composed of well known mechanisms such as the tendency to cut back safety and disaster prevention budgets in "normal" times (the "atrophy of vigilance" phenomenon (Freudenberg 1992), which is a subset of the more general "issue-attention cycle" (Downs 1972)). Only multiple action thresholds can avoid such difficulties.

Strengths and weaknesses of SHARK-type management systems

The case for a homeostatic, SPRAT-type, approach to risk management is easiest to make by comparing it with the "direct steering" approach of day-to-day hands-on political supervision of risk management issues, and showing the difficulties that such a system must face once the risks to be managed go beyond a minimal level of scale and complexity. But a more plausible alternative to SPRAT for complex public risk management, is not direct steering but SHARK, and it is SHARK that offers a more promising basis for answering the challenge issued by the champions of SPRAT, namely the proposition that "there is no workable alternative" to that model.

The general idea of a SHARK-type management system is far from new. It is commonplace in the literature of political science, in attempts to understand how real-life political systems handle policy-making tasks such as budgeting, without following the impossible precepts of fully rational decision-making. More specifically, Shrader-Frechette (1991) has developed ideas for reforming risk management (by interfering in the current legal balance between risk creators and victims, to reduce the transaction costs of victims), which involve SHARK principles. And Schwartz & Thompson (1990: 108 and passim) have developed an idea of technology assessment that employs some elements of SHARK, by arguing that juxtaposition of the rival extreme world-views identified by grid-group cultural theory will tend to improve the quality and robustness of decision-making (using the case of a lavatory rim-block product that was technically improved after exposure to "green" attack). This idea is a valuable pioneering step, although Schwartz & Thompson do not develop it beyond first principles or root it in cybernetic analysis.

The general strengths and weaknesses of managed tug-of-war control systems have been outlined and discussed by Dunsire (1993). Following his analysis, some of the strengths of SHARK are summarized in Table 9.2. Since it requires no general underlying stable consensus on goals, SHARK makes much lower demands on social rationality than SPRAT (particularly where the latter has a single threshold of safety). Since "effecting" is not fully separated from "detecting" and "directing" processes in SHARK-type institutional systems, they are is much less likely to suffer from implementation deficits (or excesses) or from atrophy effects than SPRAT. Moreover, basing risk management on pent-up opposed forces makes that process capable of being highly responsive to outside pressures. That is, familiar regulatory problems of "capture" of regulatory institutions by a narrow group, "distortion" of preferences and "groupthink", are less likely to occur when countervailing forces are deliberately juxtaposed.

But these benefits are not costless. There are no free lunches in risk management. The corresponding weaknesses of the SHARK tug-of-war model, according to Dunsire (1993), include: lack of targetability from a central point (particularly important where prompt, specific or closely aimed actions are unavoidable; ibid.: 36); its tendency to destroy the conditions of its own success over time through routinization or "observer paradoxes" (cf. Hood 1994); and its tendency to produce perverse effects in conditions of "hair-trigger sensitivity" where a whole society is divided by a single overriding cleavage (Dunsire 1993: 38).

These weaknesses are certainly not trivial. But some of them apply to all systems of control rather than to SHARK alone. In particular, self-induced decay is a problem for all control systems, and there is no reason to believe that SHARK is inherently more prone to such decay than SPRAT. And even those which do seem to be peculiar to SHARK, such as the limited targetability problem, need to be set in context. In any system where debate turns on relatively small incremental shifts in direction (as in the recurring issue of how much effort and resources should be put into dealing with the "last 10 per cent" of a problem, such as clean-up of toxic waste sites (Breyer 1993: 11ff.)), the targetability problem will be much less serious than in circumstances where quantum shifts are at issue. Moreover, as noted earlier, the other side of the "low targetability" coin is that a SHARK-type control system will typically make lower demands on social rationality than a SPRAT-type system (see Dunsire 1993: 36).

Without denying that it has limitations of the sort instanced above, two claims can be made for the SHARK model of risk management. One claim is descriptive, the other normative. Descriptively, the SHARK model seems to be typically a better account of the way many aspects of public risk management work in practice than the SPRAT model. For example, it fits the way that law courts interpolate in the struggle between zealous regulatory agencies and anti-regulatory interests, in cases such as asbestos or toxic waste (Breyer 1993: 14) and with the way that government itself interpolates between corporations, lawyers and victims, for instance in the cap imposed by the Price–Anderson Act in the USA on the liability of nuclear reactor operators (Shrader-Frechette 1991: 15). It may even fit some of the complexities of negotiating processes between regulatory agencies and other risk policy stakeholders, which have been described by sociologists such as Wynne (1987, 1992) and Irwin (1992), which are far removed from the rational decision-making aspirations of the SPRAT model. And apart from these descriptive advantages, SHARK has prescriptive advantages over SPRAT, in that it offers the basis of a management system that is both "viable" in a cybernetic sense and (unlike SPRAT) goes with, rather than against, the flow of risk management politics.

Table 9.2 Potential advantages of SHARK relative to SPRAT.

Control element	SHARK model	Advantages over SPRAT model
Director (sets goals)	No fixed targets	Makes lower demands on social rationality
Detector (checks course)	Politics and sentiment	Less vulnerable to decay/atrophy effects
Effector (changes state of system)	Tug-of-war between rival institutional pressures	Powerful servomechanism for responsiveness to outside pressures

Three general operational implications of SHARK as an institutional model of risk management

Dunsire's basic mechanical metaphor for managed competition as a system of control is the familiar type of "Anglepoise" desk lamp, whose position is controlled by counterbalancing springs. The pent-up balance of forces between the springs is such that the lamp can readily be readjusted to any point within some defined range by very light external intervention – "fingertip control" (Dunsire 1986: 344; 1992: 28–9). What is involved in such a system is not simply an equilibrium produced by uncontrollable forces in conflict – an idea present in the social sciences from their earliest beginnings – but a manageable or adjustable equilibrium. Without the underlying tension, any change would be impossible, or at least much more costly, to effect.

Dunsire's mechanical analogy is compelling, and so are alternative biological analogies for such a system. But, as noted earlier, the operational implications of designing such a system for risk management have not been investigated beyond first principles. Important unanswered questions include: what would a working model of SHARK look like? Exactly how easy is it to "engineer" such management systems in a social, rather than a mechanical, context? Can a stable system of management be built according to such principles, or does the SHARK approach imply a "footloose" and opportunistic hunt for one-off intervention niches? And just how different in practice would SHARK be from SPRAT?

There seem to be three general conditions that need to be satisfied to make the SHARK model work: a source of power in the form of forces in conflict, institutional arrangements designed to avoid peace breaking out, and a balance-tipping mechanism. These general conditions will be briefly discussed below, and the next section will consider five more specific conditions for operationalizing SHARK.

A source of power

SHARK can work only if it has a source of power in the form of immanently opposed forces that are capable of being locked into continuing conflict with one another (and can be prevented from giving up the struggle at the times when the balance tips against them). Whereas SPRAT requires a stable consensus over risk settings, built-in conflict is a prerequisite for SHARK. So, if lingering conflict is a danger sign for SPRAT, it is consensus which is the danger sign for SHARK, indicating that the policy habitat is disappearing.

Ordinarily, cultural variety and the logic of political coalition formation (in the sense of the tendency for grand coalitions to disappear) can be relied upon to provide the raw materials in the more politicized areas of risk management, such as food safety or toxic waste. But the orthodox transaction costs analysis of organized group formation (Wilson 1980: 357–74), as depicted in Table 9.3, suggests that the most favourable conditions for SHARK will arise where both the costs and the benefits of risk regulation are concentrated (e.g. where organized labour confronts organized business over safety legislation), meaning that the relative costs of group formation are low but the stakes are high. Accordingly, cell 1 in Table 9.3 provides the strongest "immanent" conditions for SHARK. Equally, cell 4, where both the costs and benefits of risk regulation are diffuse (e.g. in global environmental change) produces a policy habitat unfavourable for SHARK, and may be more suitable for SPRAT (cf. May's 1991 characterization of some aspects of risk management policy as "politics without publics").

Cells 2 and 3 are intermediate cases for SHARK. In cell 2, risk regulation benefits are diffuse, but costs are concentrated, meaning that the pro-regulation lobby will be under-organized. An example is what Shrader-Frechette (1991: 70–1) calls "the contributor's dilemma", in which risks

Table 9.3 SHARK and group formation conditions.

Costs of risk regulation	Benefits of risk regulation	
	Concentrated	Diffuse
Concentrated	(1) Most favourable conditions for SHARK Example: organized employee versus organized employer risk conflicts	(2) Beneficiary mobilization needed for SHARK Example: risks created by aggregation of individually harmless items
Diffuse	(3) Maleficiary mobilization needed for SHARK Example: risk displacement policies	(4) SHARK goes against the institutional flow Example: risks of global environmental change

Source: adapted from Wilson (1980: 357–74).

of cancer are a product of an aggregation of exposures to certain carcinogens, each of which on its own is relatively harmless. The analogy is with a small-claims problem. To set up SHARK in such circumstances requires artifice in the form of bureaucratic or policy entrepreneurship, to mobilize the pro-regulation element, for example by encouraging class action or no-win-no-fee legal practices, which lower the cost of pursuing small claims. In cell 3, it is the other way round, and here the anti-regulation element needs to be mobilized in order to make SHARK work. Examples are cases in which risks for a concentrated group can be reduced at the cost of increasing risk for a diffuse or less organized group, as in Shrader-Frechette's (1991: 67) US example of the reduction of risk for meat-cutters (by installation of guard rails around cutting machinery) at the cost of increased consumer risk from infected meat, or in the NIMBY phenomenon, in which wealthy industrial societies export toxic waste to Third World countries.

Institutionalized conflict

It follows from the analysis of Table 9.3 that for SHARK, incompatible preferences need to be institutionalized in ways that minimize collaboration or sympathy, but still keep conflict in play. Indeed, the inherent conflict need not necessarily be expressed in an overtly "warring" style, and may well involve cooperative behaviour, as long as different incompatible motives are involved in the exchange, as in the relationship between buyer and seller. This condition is more stringent that may appear at first sight. Keeping conflict institutionalized in a social system can be just as difficult as maintaining consensus. The "expert consensus" over nuclear power risks in the UK for three decades is an example of absence of conflict in the professional policy community making risk management virtually impenetrable to outside influence. Just as competitive markets can easily turn into cosy cartels, institutional systems of mutual challenge (such as professional peer-group appraisal, academic refereeing, government committees, some professional wrestling matches) can degenerate into knock-for-knock coexistence conventions or "friends-and-neighbours politics". Preventing that outcome is the major challenge to providing a habitat for the SHARK model of risk management.

A residual balance-tipping mechanism

As noted earlier, the beguiling promise of SPRAT is to avoid the ambiguities, "irrationalities" and game-playing of politics by demanding social pre-commitment to a set of rational risk acceptability standards, thereafter taking risk management "out of politics" and into the field of technical expertise. The SHARK model aims to do exactly the opposite. Although, like SPRAT, it relies heavily on servomechanisms, it is not a "look-no-hands" system of management, but a way of putting risk management more effectively into politics. Instead of assuming precommitment to the "expertized" settings of a risk thermostat, SHARK means setting up the policy machinery in such a way that politicians, public and bureaucrats are forced to confront the issues and cannot practise "management avoidance", however much they might wish to do so. SHARK does that by putting those actors in the position of balance-tippers whose weight controls the balance of pent-up opposed forces in the system. Such mechanisms are SHARK's analogues to the role of the brain in relation to nervous impulses or the water balance in the human body.

The process of balance-tipping under SHARK could, in principle, operate in several ways, and is not restricted to any single vision of good government. If good government is conceived as strong representative democracy, then the institutions and players who have the key role of "balance-tippers" will be ministers and legislators. But SHARK is also compatible with less orthodox doctrines of good government, such as direct democracy (a feature of Ostrom's 1974 vision of "democratic administration") or even the classical democratic idea of selection of public officials by lot (as in Burnheim's 1985 recipe for "demarchy", government by committees of randomly selected citizens). In such cases, the role of selectively inhibiting the institutionalized contestants in the risk policy tug-of-war would belong instead to law courts and juries, to tribunals of public opinion such as referenda or polls (Bentham's idea of a "public opinion tribunal", as reflected today in experiments with "deliberative polling"), or to special forums such as science courts, as advocated by Majone (1989). Like SPRAT, SHARK is adaptable to any of those visions of good government. But, unlike SPRAT, it will also fit with a conception of "no-one in charge management" (Bryson & Crosby 1992) in which the role of balance-tipper is ill defined and evanescent.

Five specific conditions for the SHARK strategy

The discussion above has mingled the terminology of "design" and "habitat", suggesting two general ways in which SHARK can be made to work. One is an opportunistic "biological niche" approach, which focuses on niches for intervention "wherever the action is". The SHARK strategy for risk managers (however conceived) is one of opportunistic and shifting intervention, seeking "making extremes meet" social conditions wherever they are to be found. The policy skills involved are those of identifying a niche or a lever for influence in a structure of conflict. Shrader-Frechette's (1991: 197–218) proposals for procedural reforms in risk management to alter the balance between risk creators and risk receivers (by means such as changing the burden of evidence required in litigation from toxic tort victims, encouraging class action suits, provision of public funding to ensure equal access to technical expertise in negotiations over risk issues between citizens and business groups, and the institution of adversary proceedings on a "science court" basis, involving scientists and lay-people) are examples of such niche-spotting policy skills. Indeed, seen in this way, SHARK need not involve any permanent institutions or arrangements, but may operate as a way of shifting the "action" around the social landscape, with public funding or other resources moving from one fulcrum to another according to the conditions of the moment.

The other possible approach to making SHARK work is much more demanding. That is to set up SHARK like a physical design artefact, as a more or less permanent set of institutional arrangements. Such an approach would be revolutionary, in the sense that it would require major demolition work on current risk management institutions, particularly in state bureaucracies, and the rebuilding of those institutions according to a very different blueprint. To redesign risk management bureaucracy so that its settings can be "turned on a sixpence", providing the conditions for "fingertip control" like an "Anglepoise" lamp, would require the prior use of power on a considerable scale. Yet everything social scientists know about organization suggests that it is, if anything, even harder to reconstruct existing institutions all at once than it is to achieve social consensus on acceptable risk.

If that is true, a viable strategy for SHARK would seem to imply something much more like the first than the second approach – niche-finding, not machine-building. It would need to involve a set of incremental adaptations of what already exists, building on existing "materials", such as conflicts between budget allocators and applicants, auditors and the audited, producers and consumers, managers and professionals, plaintiffs and defendants. Five possible institutional arrangements which can be found (or built up by degrees) to strengthen SHARK are briefly discussed below.

Encouraging explicit institutionalization of rival values

SHARK is likely to work best in conditions where there are institutionalized champions for each of the rival values (such as aggregate economic growth, public health, individual freedom), which clash in public risk management. The policy implication is that risk management responsibilities in state bureaucracy should be divided by value, not by function alone (as in the conventional style), to make the risk management trade-offs explicit and transparent.

The idea of organizing around values rather than functional responsibilities is not new. Some years ago; for example, the New Zealand Treasury (1987: 78–9) – no doubt with "making extremes meet" control in mind – proposed that Cabinet portfolios should be distributed on this principle. They argued that for government to make decisions in a way that most closely reflected the preferences of the electorate as a whole required:

> . . . the elucidation of conflict between broadly defined values – those fundamental goals of any society which must be traded off; that is, the goals to which it is the Government's responsibility to assign weights reflecting the preference of the electorate. Under this schema, for example, one could postulate a policy advisory agency concerned with the goal of maximizing the wealth of the economy, analyzing policy from the perspective of its impact on growth; another ministry could be given responsibility for looking at policy from the perspective of equity – how is any given level of income to be shared among participants in the economy and how will particular policies impact on the distribution of income? Advice provided on such a basis would make explicit, at a very fundamental level, the trade-offs the Government must make . . .

To reorganize state risk management bureaucracies on such lines, at least two critical difficulties would need to be solved. One is how to identify and represent the different values in the field, and how to assign weight to them. The other is how far to take the principle of rebuilding state bureaucracies around a single value, particularly where technical knowledge is in short supply, as in areas of high-technology risk such as nuclear power (ibid.: 79). Such difficulties are far from trivial. But many value conflicts are already built into existing institutional arrangements (often in an unacknowledged way), and extending that design principle at the margin (as opposed to once-for-all transformation) is far from inconceivable.

What are the relevant values to be separated and juxtaposed? The conflict between aggregate wealth maximization and distributional equality pervades much of risk management policy, and SHARK would imply that such conflicts should not be internalized in a single institutional unit.

A second value-conflict, clearly demonstrated in the debate over regulation of genetically engineered organisms, is the clash between the *Versorgensprinzip* (the precept of "stopping things before they start", even before "harm" is established beyond a peradventure by orthodox canons of science) and the rival doctrine of "resilience" (Wildavsky 1988), that is, the precept of maintaining social capacity to learn from experience and to cope with the unexpected. A third conflict is between the avoidance of Type I and Type II errors, a pervasive problem in risk management (Raiffa 1968). The danger of creating harm by policy inaction (e.g. when lives are lost as a result of failure to prohibit a dangerous drug or evacuate a dangerous area) needs to be weighed against the opposite danger of creating harm by policy intervention. Examples of the second type of harm include loss of life or other harm arising from an evacuation that turns out to be unnecessary, where "risk compensation" responses to regulation produce risk-shifting, for example from drivers to pedestrians (Adams 1985) or even to increased risk, as in those cases where mandatory child-proof aspirin bottle tops cause adult users simply to leave the tops off, making children more rather than less vulnerable (Breyer 1993: 10).

Separate institutionalization of these three value-conflicts alone would go a long way to making SHARK a reality. And such institutionalization does not necessarily require reconstruction of Cabinet government from the ground upwards, because rival values can be institutionalized in different ways and at different levels – for example, advisory or executive organization, independent regulatory agencies or core public bureaucracies, divisions within organizations or corporate entities. Moreover, SHARK does not necessarily require exact institutional balance of the rival risk management values, and could develop by an iterative process of identifying the value trade-offs, exploiting the raw materials that are already to hand.

Developing incentive structures which reward confrontation and encourage value champions to take extreme positions

The logic of the SHARK risk management model does not stop with the implication that the different values in play should be institutionalized. It also implies that there are advantages in those different institutional units being separated, to limit "fraternization with the enemy". The analogy is with the way that monarchs in former times often forbade the members of their courts to meet except in the royal presence, for fear of collaboration behind their backs, or with the more modern and widespread practice (both in public and private management) of "Chinese walls" separating different organizational functions, often underpinned by separate organizational structures, recruitment and career paths and expressed in devices such as two-key or dual-signature authorization systems.

The next stage beyond mere separation of functions is the development of positive ex officio antagonisms through role antonyms (following the model of landlord/tenant, producer/consumer or prosecution/defence). Well known institutional devices in risk management that reflect this principle are statutory requirements for the existence of positions such as safety officers or committees, coupled with the prescription that particular kinds of information are conveyed to company boards. Such role-antonyms can help to ensure that the rival champions have as much incentive as possible to "go for the jugular" in their struggles against their opponents, rather than going for the "middle ground" in the classic British civil service fashion. Open adversarial exchanges, to be discussed below, will ordinarily be required to sustain such incentives and maintain the mutual repulsion of the combatants.

Develop schizophrenic incentive structures (second chances for failures or "Geneva conventions on shooting survivors")

Although SHARK needs to be built on role antagonisms with strong incentives to promote rival values, such a system will quickly self-destruct unless there are also obstacles in the way of once-for-all victory for any one of the warring parties. The more difficult it is for any one of the rival players in the policy community to "go out of business" completely, the longer SHARK can survive beyond a single policy correction. Institutional mechanisms for "wiping the slate clean" after each bout or round in the struggle, in terms of the position and resource base of rival value-guardians, can help to fulfil this condition.

Such mechanisms have developed in other domains. Examples of institutionalized ways of keeping losers in business are the "second chances for failures" that conventional institutions of bankruptcy offer in market capitalism (thereby underpinning the forces of entrepreneurship and competition at the expense of absolute integrity of the credit system) and the "Geneva conventions" on shooting survivors in warfare. A version of SHARK that was capable of surviving institutionally over several policy corrections would need a way of keeping losers in business without removing the will to win, implying the same sort of schizophrenic incentive structure observable in the rules of war and insolvency.

Regular (but not continuous) adversarial exchanges

SHARK is less likely to break down through drift towards a "middle ground" if there are regular adversarial forums in which the risk management debate is conducted. Public forums encourage conflict rather than cooperation, require the rival players to show their mettle, and help to equalize the balance between big and small actors by substituting political or intellectual methods of resolution (such as majority rule) for other forms of clout. Making such adversarial forums regular, but not continuous, helps to sustain the sense of occasion and increases the number of winnable encounters.

Adversarial forums (such as court cases, committee hearings, planning inquiries) are commonplace in institutional life (cf. Bryson & Crosby 1992). As noted earlier, there have been many advocates for "science courts" constituted along analogous lines for handling the "trans-scientific" aspects of risk regulation, such as extrapolation from high dose–response observations to low, unobservable levels (cf. Majone 1989). Such proposals reflect the logic of SHARK, but such forums can obviously be distorted by elements such as scarcity and uneven distribution of the key technical knowledge available, or domination by the best-funded or most tightly organized interests rather than by the most convincing arguments. To avoid such shortcomings, there is a need for procedural arrangements to be designed to ensure a high premium on technical argument (or scientific proceduralism, as Shrader-Frechette (1991) calls it) and methods for avoiding staleness, such as turnover of personnel.

Encourage caucus race resourcing

In the human body's nervous system, a "victory" to one side (through panic activity or torpidity) does not starve the other side of resources to carry on the fight in the future. In the same way, SHARK is most likely to be sustained in conditions where resources can be distributed in such a way that the various rival players stay in the game, without destroying the incentive to take conflicts "down to the wire". That is, the resourcing of the those players would need to resemble the "caucus race" in Lewis Carroll's *Alice in Wonderland* – a race in which all the contestants won prizes.

The best-known institutional arrangements for resourcing all legitimate comers in a regulatory system is the "subsidiarity" doctrine, which has traditionally been used for public service delivery in many of the communally divided societies of Continental Europe. The subsidiarity doctrine does not simply mean that different social groups are entitled to participate in public service provision so long as they are capable of doing so. It also means that such groups are entitled to support from the state for so doing (cf. Hood & Schuppert 1988: 19). Equal funding arrangements for public campaign groups engaged in risk policy (to prevent "capture" of public bureaucracies or walk-over victories for the best-funded organizations) thus seem to be a central requirement for the development of SHARK.

Resources are not necessarily to be equated with cash; and indeed an alternative way of structuring a "caucus race" is for government to promote the transfer of talent among different institutions, which does not necessarily undermine the "will to win" among participants in the same way as strict fiscal equalization would do. A classic case of such a transfer is the shift in 1957 of Lord (Christopher) Hinton, one of the UK's early "nuclear knights", from the position of Managing Director of the Industrial Group of the Atomic Energy Authority (then the key producer in the civil nuclear power field) to the chairmanship of the then Central Electricity Generating Board (the main purchaser of nuclear power plants in the UK). A classic SHARK strategy, this shift meant that the CEGB for the first time gained the expertise and inside knowledge to challenge the promoters of the nuclear power programme on their own technical ground (Williams 1980: 21, 25). Breyer (1993: 60–1) sees the movement of career professionals as the central issue for reform of risk management in the USA, and advocates the creation of a *cursus honorum* (elite career structure) for US risk regulators through health and environmental agencies, Congress and Office of Management and Budget, with the aim of using planned career paths to create greater coherence and uniformity in risk management. From a SHARK perspective, a similar strategy is appropriate – but to sharpen the clash of institutional opposites rather than to promote consensus.

As noted earlier, the advantage of equalizing through people rather than through finance is that such transfers do not undermine the protagonists' will to win (just as professional football players can be motivated to strive as hard as they can for the team they happen to be with at any time, even though their very success in doing so may mean they later find themselves courted to play for an opposing team) and that they cause detailed knowledge to be carried from one group to its rivals, thereby enabling more effective counterpunching from better knowledge of the detailed positions of its opponents. Of course, such a strategy excludes those "players" – far more common in risk management than in football – who operate out of principled conviction, rather than functioning as performing professionals available for hire to anyone who will employ them. But such defections do nevertheless happen, even in such "ideological" or politicized fields of public policy as gun control (cf. Wright 1988), and their existence, or possibility, is a key resource for SHARK.

Conclusion

Excessive reliance on dichotomies is dangerous. Real-life risk management will always be some mixture of basic styles rather than any one pure type. Current practice is a mixture of SHARK and SPRAT, and is sometimes a case of the latter masquerading as the former. Moreover, human control systems in risk management, as in other fields, tend to change over time, both because of an internal dynamic – control systems tend to "wear out" over time, undermining their own bases – and because their social habitat alters. Indeed, perhaps the basic problem with SPRAT as a recipe is not so much its technical defects (substantial as they are) as the fact that it fits poorly with the general move away from "hierarchist" attitudes that cultural theorists such as Douglas (1990: 12–3) observe in contemporary capitalism, and the consequently increasing politicization of risk management between individualist and fundamentalist world-views. Waning trust in authority and conventional science increasingly weaken the social foundations of that model.

The argument of this contribution is that, although SPRAT is always likely to have a place in risk management, there is no reason why it should be the only available approach. *Pace* the "enlightened engineers", there is a viable alternative to the SPRAT recipe. SHARK offers a different way of doing public risk management precisely in those policy habitats in which it is difficult for SPRAT to survive and flourish; notably circumstances of "trans-scientific" knowledge, low trust and high politicization.

The general control-theory logic of SHARK is far from new, having been spelt out by Dunsire (1978) nearly two decades ago. And, as has been noted above, many of the specific measures espoused by risk management reformers in recent years, such as Shrader-Frechette (1991), follow SHARK principles. All this contribution has tried to do is to bring those two elements together and to spell out some of the conditions needed to operationalize SHARK as a generalized approach to risk management. Drawing on the discussion in the last section, Table 9.4 summarizes some of the main institutional features that differentiate SPRAT and SHARK, distinguishing the implications for the organization of government bureaucracy, for regulatory strategy, for the organization of decision advice procedures and for the direction of public funding.

Table 9.4 Institutional features of SPRAT and SHARK.

Institutional dimension	SPRAT style	SHARK style
Method of organizing government bureaucracy	Organized by functional policy area	Organized by underlying value
Main strategy for regulation of business and other organizations	Stress placed on specifying substantive outcomes and technical standards	Stress placed on specifying decision procedures and developing role-antonyms
Decision advice procedures	Stress placed on consensus-seeking forums	Stress placed on conflict-seeking adversarial forums
Main strategy for allocation of public funds	Funds targeted on established expertise and authority	Funds available to all legitimate comers

Though Dunsire (1978) originally wrote of a system of managed competition as a relatively stable and permanent device for keeping public bureaucracies under control, his later writings (1990, 1992, 1993) have moved away from institutional design in that sense, to portraying managed competition more as an ephemeral (but highly effective) basis for one-shot policy interventions. Certainly, the discussion in the previous two sections suggests that SHARK is institutionally quite complex to "engineer" as a semi-permanent management control system and is likely to be far from "free" in terms of resources. Ironically perhaps, making extremes meet institutionally seems to demand several very delicate balancing acts in order to create incentive structures that would make the model work properly. SHARK is not a panacea, and cannot be applied by rule and rote.

It follows that SHARK may normally be better conceived in biological rather than engineering terms, in the sense that the conditions for its operation need to be "found" rather than "made" – to be spotted in a potential, rather than realized, form and then developed. But, if the analysis of Table 9.3 is correct, there is likely to be scope for such discoveries, since only in the "majoritarian politics" conditions of cell 4 of that table are the raw materials for SHARK likely to be completely absent. Indeed, the SHARK-managed competition model is strong in the very circumstances where the SPRAT model hits trouble – highly politicized risk management, where experts disagree and trust in expertise and authority is low or polarized. The great advantage of the SHARK model is that it fits with the confrontational dynamic of risk politics rather than trying to work against it.

BIBLIOGRAPHY

Ackoff, R. 1970. *A concept of corporate planning*. New York: John Wiley.
ACSNI (Advisory Committee on the Safety of Nuclear Installation Study Group on Human Factors) 1991. *Second Report: human reliability assessment – a critical overview*. London: HMSO.
— 1993. *Advisory Committee on the Safety of Nuclear Installations: Human Factors Study Group, Third Report*. London: HMSO.
Adams, J. G. II. 1985. *Risk and freedom: the record of road safety regulation*. London: Transport Publishing Projects.
— 1995. *Risk*. London: IICL Press.
Adams, J. G. II. & M. Thompson 1991. *Risk review: perception varieties of uncertainty, sources of information*. End of award report, ESRC, Swindon.
Aggarwal, R. & L. Soenen 1989. Project exit value as a measure of flexibility. *The Engineering Economist* **35**, 39–54.
Aircraft Accident Report 79–139, 1980. *Aircraft Accident Report 79–139; Air New Zealand McDonnell–Douglas DC-10 ZK-NZP Ross Island, Antarctica, 28 November 1979*. Wellington: Office of Air Accidents Investigation, Ministry of Transport.
Alexander, D. 1993. *Natural disasters*. London: IICL Press.
Alicke, M. D. 1992. Culpable causation. *Journal of Personality and Social Psychology* **63**, 368–78.
Allaire, Y. & M. Firsitotu 1989. Coping with strategic uncertainty. *Sloan Management Review* **30**, 7–15.
Amendola, A., S. Contini, I. Ziomas 1992. Uncertainties in chemical risk assessment: results of a European benchmark exercise. *Journal of Hazardous Materials* **29**, 347–63.
Ansell, J. & F. Wharton (eds) 1992. *Risk analysis assessment and management*. Chichester: John Wiley.
Appleby, P. 1949. *Policy and administration*. Tuscaloosa: Alabama University Press.
Argenti, J. 1976. *Corporate collapse*. New York: McGraw-Hill.
Aronowitz, S. 1988. *Science as power: discourse and ideology in modern society*. Minneapolis: University of Minnesota Press.
Ashby, W. R. 1956. *An introduction to cybernetics*. London: Chapman & Hall.
Ashworth, A. 1992. *Sentencing and criminal justice*. London: Butterworth.

Baldissera, A. 1987. Some organisational determinants of technological accidents. *Quaderni di Sociologia* **33**(8), 49–73.
Baram, M. 1991. Rights and duties concerning the availability of environmental risk information to the public. In *Communicating risks to the public*, R. E. Kasperson & P. J. M. Stallen (eds), 67–78. Dordrecht: Kluwer.
Bardach, E. & R. A. Kagan 1982. *Going by the book: the problems of regulatory unreasonableness*. Philadelphia: Temple University Press.
Barnaby, K. C. 1968. *Some ship disasters and their causes*. London: Hutchinson.
Bauman, Z. 1993. *Postmodern ethics*. Oxford: Basil Blackwell.
Beaty, D. 1991. *The naked pilot*. London: Methuen.
Beck, U. 1992. *Risk society: towards a new modernity* [translated by Mark Ritter]. London: Sage.
Beder, S. 1991. The fallible engineer. *New Scientist* (2 November), 38–42.

Beer, S. 1966. *Decision and control*. London: John Wiley.

Bensman J. & I. Gerver 1963. Crime and punishment in the factory: the function of deviancy in maintaining the social system. *American Sociological Review* **28**(4), 588–98.

Bergman, D. 1991. *Deaths at work: accidents or corporate crime*. London: Workers' Educational Association.

Bergman, D. 1994. *The perfect crime?* Birmingham: West Midlands Health and Safety Advice Centre.

Bertalanffy, L. von 1968. *General systems theory*. New York: George Braziller.

Best, G., G. Parston, J. Rosenhead 1986. Robustness in practice – regional planning of health services. *Journal of the Operational Research Society* **37**, 463–78.

Bijker, W. E. & J. Law 1992. *Shaping technology/building society: studies in sociotechnical change*. Cambridge, Massachusetts: MIT Press.

Blaikie, P., T. Cannon, I. Davis, B. Wisner 1994. *At risk: natural hazards, people's vulnerability and disasters*. London: Routledge.

Blockley, D. I. 1985. Reliability or responsibility. *International Journal of Structural Safety* **2**, 273–80.

—1991. Hazard engineering and management. Paper presented at the ESRC/LSE Seminar Series, London School of Economics and Political Science, London, May.

— 1992a. *Engineering safety*. Maidenhead: McGraw-Hill.

— 1992b. Engineering from reflective practice. *Research and Engineering Design* **4**, 13–22.

Bohm, D. 1980. *Wholeness and the implicate order*. London: Routledge & Kegan Paul.

Bowman, E. H. & H. Kunreuther 1988. Post-Bhopal behaviour at a chemical company. *Journal of Management Studies* **25**, 387–402.

Braithwaite, J. 1991. Shame and modernity. *British Journal of Criminology* **33**, 1–18.

Breyer, S. G. 1993. *Breaking the vicious circle: toward effective risk regulation*. Cambridge, Massachusetts: Harvard University Press.

Brogan, A. J. 1991. Safety case and/or rules: substitute or complementary? A vision of the future offshore safety regime? *Det Norske Versitas Classification* (unpublished).

Brown, P. (Phil) 1987. Popular epidemiology: community response to toxic waste-induced disease in Woburn, Massachusetts. *Science, Technology and Human Values* **12**, 78–85.

— 1993b. Popular epidemiology challenges the system. *Environment* **35**(8), 17–41.

Brown, P. (Phyllida) 1993a. The hour of the activist. *New Scientist* (3 April), 14–15.

Bryant, E. A. 1991. *Natural hazards*. Cambridge: Cambridge University Press.

Bryson, J. & B. Crosby 1992. *Leadership for the common good: tackling public problems in a shared-power world*. San Francisco: Jossey–Bass.

Bullard, D. 1994. Overcoming racism in environmental decision-making. *Environment* **36**(11), 10–20, 39–44.

Burnheim, J. 1985. *Is democracy possible? The alternative to electoral politics*. Cambridge: Polity.

Burton, I., R. W. Kates, G. F. White 1978. *The environment as hazard*. New York: Oxford University Press.

—1993. *The environment as hazard*, 2nd edn. New York: Guilford Press.

Bush, R. 1986. Between two worlds: the shift from individual to group responsibility in the law of causation of injury. *University of California Law Review*, 1473.

Butterworth, E. & D. T. H. Weir 1990a. Social and management problems of disasters. Paper presented at Annual Study of the Institute for Civil Defence, 1990.

— 1990b. Causes of failure in complex systems. Paper presented at Annual Conference of Royal Institute for Public Administration, 1990.

Cacciabua, P. C., I. Gerbaulet, N. Mitchison (eds) 1994. *Safety management systems in the process industry*. EUR 15743 EN CEC. Luxembourg: Joint Research Centre, European Commission.

Cairncross, F. 1991. *Costing the earth*. London: Business Books/The Economist.

Calabresi, G. 1970. *The costs of accidents*. New Haven, Connecticut: Yale University Press.

Calavita, K., J. Dimento, G. Geis, G. Forti 1991. Dam disasters and Durkheim: an analysis of the theme of repressive and restitutive law. *International Journal of Sociology of Law* **19**, 407–26.

Carlsson, B. 1989. Flexibility and the theory of the firm. *International Journal of Industrial Organisation* **7**, 179–203.

Carson, W. 1982. *The other price of Britain's oil*. Oxford: Martin Robertson.

Carter, N. & P. Lowe 1994. Environmental politics and administrative reform. *The Political Quarterly* **63**(3), 263–74.

CBI 1986. *Clean up – it's good business*. London: Confederation of British Industry.

—1990. *Developing a safety culture: business for safety*. London: Confederation of British Industry.

CEGB 1982. *Proof of Evidence, Sizewell B Power Station Public Inquiry: Proof 2 and 4*. London: Central Electricity Generating Board.

Chalmers, A. 1990. *Science and its fabrication*. Minneapolis: University of Minnesota Press.

Chappell, S. 1994. Using voluntary incident reports for human factors evaluations. In *Aviation psychology in practice*, A. N. Johnston, N. J. McDonald, R. G. Fuller (eds), 149–69. Aldershot: Avebury Technical.

Checkland P. 1981. *Systems thinking systems practice*. London: John Wiley.

Chemical Industries Association 1987. *A guide to hazard and operability studies*. London: Chemical Industry Safety, Health & Environment Council.

Coker, A. M. & C. Richards 1992. *Valuing the environment: economic approaches to environmental evaluation*. London: Pinter (Belhaven).

Collingridge, D. 1980. *The social control of technology*. Milton Keynes: Open University Press.

—1983. *Technology in the policy process: controlling nuclear power*. London: Pinter.

—1992. *The management of scale: big organizations, big decisions, big mistakes*. London: Routledge.

Collingridge, D. & P. James 1991. Inflexible energy technologies in a rapidly changing environment. *Long Range Planning* **24**, 101–7.

Courtis, J. 1986. *The 44 most common management mistakes and how to avoid them*. London: British Institute of Management and Kogan Page.

Cullen, The Honourable Lord 1990. *The public inquiry into the Piper Alpha disaster*. London: HMSO.

Cuny, F. C. 1983. *Disasters and development*. Oxford: Oxford University Press.

Davies, P. 1987. *The cosmic blueprint: new discoveries in Nature's creative ability to order the universe*. New York: Touchstone Press.

Deansley, C. & C. Papanicolaou 1992. The big clean-up begins. *Post Magazine* [risk management supplement] (Spring).

Degg, M. 1992a. The ROA earthquake hazard atlas project: recent work from the Middle East. In McCall et al. (1991: 93–104).

—1992b. Some implications of the 1985 Mexican earthquake for hazard assessment. In McCall et al. (1991: 105–114).

Degg, M. & J. C. Doornkamp 1989. *Earthquake hazard atlas, 1: Israel*. London: Reinsurance Offices Association.

—1990. *Earthquake hazard atlas, 2: Egypt*. London: Reinsurance Offices Association.

—1994. *Earthquakes* [Reinsurance Technical Report, January]. London: Reinsurance Offices Association.

Denison, D. 1984. Bringing corporate culture to the bottom line. *Journal of Organisational Dynamics*, 5–22.

Denning, Lord 1978. *Alidair vs Taylor. Industrial Cases Reports*, 445–6. London: Council of Law Reporting.

Department of Transport 1990. *Report on the collision that occurred on 4th March 1989 at*

Purley. London: HMSO.

Dester W. 1992. *The development of a structure for hazard audits.* PhD thesis, Department of Civil Engineering, University of Bristol.

Dietz, T. M. & R. W. Rycroft 1987. *The risk professionals* New York: Russell Sage Foundation.

Dixon, N. 1976. *On the psychology of military incompetence.* London: Jonathan Cape.

Douglas, J. 1995. Disaster preparedness and mitigation: the UK approach. In *Windstorm*, 60–66. London: The Royal Academy of Engineering.

Douglas, M. 1985. *Risk acceptability according to the social sciences* New York: Russell Sage Foundation.

—1986. *Risk acceptability according to the social sciences.* London: Routledge.

—1987. *How institutions think.* London: Routledge.

—1990. Risk as a forensic resource. *Daedalus* [Journal of the American Academy of Arts and Sciences] **119**(4), 1–16.

—1992. *Risk and blame: essays in cultural theory.* London: Routledge.

—1994. The politicisation of risk and the neutralisation of politics. Paper presented to the Political Economy Research Centre seminar, University of Sheffield, 17 February.

Douglas, M. & A. Wildavsky 1982. *Risk and culture: an analysis of the selection of technological dangers.* Berkeley: University of California Press.

Downs, A. 1972. Up and down with ecology: the issue-attention cycle. *Public Interest* **28**(1), 38–50.

Drabek, T. E. 1986. *Human system responses in disaster: an inventory of sociological findings.* New York: Springer.

Drabek, T. E. & E. L. Quarantelli 1967. Scapegoats, villains and disasters. *Trans-Action* **4**, 12–17.

DTI 1991. *The quality gurus.* London: Department of Trade and Industry.

Dunsire, A. 1978. *Control in a bureaucracy: the execution process*, vol. 2. Oxford: Martin Robertson.

—1986. A cybernetic view of guidance, control and evaluation in the public sector? In *Guidance, control and evaluation in the public sector*, F-X. Kaufman, G. Majone, V. Ostrom (eds), 327–46. Berlin: de Gruyter.

—1990. Holistic governance. *Public Policy and Administration* **5**(1), 4–19.

—1992. Modes of governance. In *Modern governance*, J. Kooiman (ed.), 21–34. London: Sage.

—1993. *Manipulating social tensions: collibration as an alternative mode of government intervention.* Discussion Paper 13/7, Max-Planck-Institut für Gesellschaftsforschung, Cologne.

Dynes, R. R., B. De Marchi, C. Pelanda (eds) 1987. *Sociology of disasters: contribution of sociology to disaster research.* Milan: Franco Angeli.

The Economist 1994. At risk: a survey of insurance. (3 December)

Einhorn, H. J. 1986. Accepting error to make less error. *Journal of Personality Assessment* **50**(3), 387–95.

Eiser, R. E. 1994. *Attitudes, chaos and the connectionist mind.* Oxford: Oxford University Press.

Elms, D. G. & C. J. Turkstra 1992. A critique of reliability. In *Engineering safety*, D. Blockley (ed.), 427–45. London: McGraw-Hill.

Eppink, D. 1978. Planning for strategic flexibility. *Long Range Planning* **11**, 9–15.

European Construction Institute 1992. *Total project management of construction safety, health and environment.* London: Thomas Telford.

Evans, A. 1992. Public transport and road safety. Inaugural lecture as London Transport Professor of Transport Safety, University College London.

Evans, A. W. & D. J. Maidment (eds) 1995. *Value for money in transport safety measures.* London: University of London Centre for Transport Studies.

Evans-Pritchard, E. 1937. *Witchcraft, oracles and magic among the Azande.* Oxford: Oxford University Press.

Evening Standard 1990. Jailing bad train drivers [editorial]. (4 September).

Federal Aviation Administration, 1992. *Air carrier voluntary disclosure reporting procedures* [Advisory Circular 120–56]. Washington DC.

Fennell, D. 1988. *Investigation into the King's Cross Underground fire*. London: HMSO.

Feyerabend, P. K. 1975, *Against method*. London: New Left Books.

Fido, A. T. & D. O. Wood 1989. *Safety management systems*. London: Further Education Unit.

Field, S. & N. Jorg 1991. Corporate liability and manslaughter – should we be going Dutch? *Criminal Law Review* (March), 156–71.

Financial Times 1994. Quality under fire. (21 June).

Fiorina, M. 1982. Legislative choice of regulatory forms: legal process or administrative process? *Public Choice* **39**, 33–66.

Fiorino, D. J. 1990. Citizen participation and environmental risk: a survey of institutional mechanisms. *Science, Technology and Human Values* **15**, 226–43.

Fischhoff, B. 1990. Psychology and public policy: tool or toolmaker? *American Psychologist* **45**, 647–53.

Fischhoff, B., S. Lichtenstein, P. Slovic, S. L. Derby, R. L. Keeney 1981. *Acceptable risk*. Cambridge: Cambridge University Press.

Fischhoff, B., N. F. Pidgeon, S. T. Fiske 1983. Social science and arms race politics. *Journal of Social Issues* **39**(1), 161–80.

Fischhoff, B. & O. Svenson 1988. Perceived risks of radionuclides: understanding public understanding. In *Radionuclides in the food chain*, M. Carter (ed.). New York: Springer.

Fiske, S. & S. Taylor 1984. *Social cognition*. New York: Random House.

Fisse, B. & J. Braithwaite 1988. Accountability and the control of corporate crime. In *Understanding crime and criminal justice*, M. Findlay & R. Hogg (eds). Sydney: The Law Book Company.

Fitzgerald, P. 1968. Voluntary and involuntary acts. In *The philosophy of action*, A. White (ed.), 120–43. Oxford: Oxford University Press.

Flight International 1991. Error and punishment [editorial]. (8 May, 3).

Fordham, M, S. M. Tunstall, E. C. Penning-Rowsell 1989. *Choice and preference in the Thames floodplain: the beginnings of a participatory approach. Landscape and Urban Planning* **20**, 183–7.

Fortune, J. & G. Peters 1995. *Learning from failure: the systems approach*. Chichester: John Wiley.

Forward, G., D. Beach, D. Gray, J. Quick 1991. Memtofacturing: a vision for American industrial excellence. *The Executive* **5**(3), 32–44.

Foster, H. D. 1980. *Disaster planning*. New York: Springer.

Freudenburg, W. R. 1988. Perceived risk, real risk: social science and the art of probabilistic risk assessment. *Science* **242**(October), 44–9.

— 1992. Nothing recedes like success? Risk analysis and the organizational amplification of risks. *Risk – Issues in Health and Safety* (Winter), 1–35.

Friend, J. & A. Hickling 1987. *Planning under pressure: the strategic choice approach*. Oxford: Pergamon.

Fujita, T. T. 1987. *US tornadoes, part 1: 70-year statistics*. Satellite and Mesometeorology Research Project, Department of Geophysical Sciences, University of Chicago.

Funtowicz, S. O. & J. R. Ravetz 1990. *Uncertainty and quality in science for policy*. Dordrecht: Kluwer.

— 1991. A new scientific methodology for global environmental issues. In *Ecological economics*, R. Constanza (ed.), 137–52. New York: Columbia University Press.

— 1992. Three types of risk assessment and the emergence of post-normal science. In *Social theories of risk*, D. Golding & S. Krimsky (eds), 251–75. West Point, Connecticut: Praeger.

Furmston, M. P. 1992. Reliability and the law. In *Engineering safety*, D. Blockley (ed.), 385–401. Maidenhead: McGraw-Hill.

Galanter, M. 1992. Law abounding: legalisation around the North Atlantic. *Modern Law Review* **55**, 1–24.

Garland, D. 1990. *Punishment and modern society*. Oxford: Oxford University Press.

Gherardi, S. & B. A. Turner 1988. *Real men don't collect soft data*. Quaderno 13, Dipartimento di Politica Sociale, Universita di Trento.

Giddens, A. 1990. *The consequences of modernity*. Cambridge: Polity.

—1991. *Modernity and self-identity*. Cambridge: Polity.

Gorz, A. 1989. *Critique of economic reason*. London: Verso.

Grayson, D. 1989. *Terror in the skies*. London: W. H. Allen.

Green, C. H., S. M. Tunstall, M. Fordham 1991. The risks from flooding: which risk and whose perception? *Disasters* **15**(3), 227–36.

Groenewold, P. & P. Vergragt 1991. Environmental issues as threats and opportunities for technological innovation. *Technology Analysis and Strategic Management* **3**(1), 43–55.

Grose, V. L. 1987. *Managing risk: systematic loss prevention for executives*. Englewood Cliffs, New Jersey: Prentice Hall.

Gross, P. R. & N. Levitt 1994. *Higher superstition: the academic left and its quarrels with science*. Baltimore: Johns Hopkins University Press.

Grove-White, R., S. Kapitza, V. Shiva 1992. Public awareness, science and the environment. In *An agenda for science for environment and development into the 21st century*, J. Dooge et al. (eds), 239–48. Cambridge: Cambridge University Press.

Guardian 1989a. "Enterprise" risk to public safety. (4 May)

—1989b. DPP to decide on manslaughter charge for BR within weeks. (8 November)

Gupta, S. & J. Rosenhead 1968. Robustness in sequential investment decisions. *Management Science* **15**, 18–29.

Hacking, I. 1986. Culpable ignorance of interference effects. In *Values at risk*, D. MacLean (ed.), 136–54. Totowa, New Jersey: Rowman & Allanheld.

Hadden, S. G. 1989. *A citizen's right to know: risk communication and public policy*. Boulder, Colorado: Westview.

Hadfield, P. 1992. Metal rod plunges reactor into chaos. *New Scientist* (10th October).

Hale, A. 1990. The human element in disasters. In *Disaster prevention and limitation*, A. Z. Keller & H. Wilson (eds), 141–53. London: British Library.

Halpern, J. J. 1989. Cognitive factors influencing decision-making in a highly reliable organization. *Industrial Crisis Quarterly* **3**(2), 143–58.

Hamer, M. 1995. Safety signals set at danger. *New Scientist* **147**(1986: 15 July), 14–15.

Hamilton, V. L. & J. Sanders 1992. Responsibility and risk in organizational crimes of obedience. *Research in Organizational Behaviour* **14**, 49–90.

Handmer, J. W. 1990. *Flood insurance and relief in the US and Britain*. Working Paper 68, Natural Hazards Research and Applications Information Center, Institute of Behavioral Sciences, University of Colorado.

—1992. Hazard management in Britain: another disastrous decade? *Area* **24**(2), 113–22.

—1995. Cooperation, coercion, capacity and commitment: key concepts in floodplain management. In *Flood protection of towns – ideas and experience* [conference proceedings], v11–v20. Cracow, Poland: Ministry of Environmental Protection, Natural Resources and Forestry.

Handmer, J. W. & E. C. Penning-Rowsell (eds) 1990. *Hazards and the communication of risk*. Aldershot: Gower.

Harle, P. 1994. Investigation of human factors: the link to accident prevention. In *Aviation psychology in practice*, A. N. Johnston, N. J. McDonald, R. G. Fuller (eds). Aldershot: Avebury.

Hay, K. 1991. Is nuclear power over-engineered? *Nuclear Engineering International* (July), 37–9.

Hewitt, K. 1983. The idea of calamity in a technocratic age. In *Interpretations of calamity*,

234

K. Hewitt (ed.), 3–32. London: Allen & Unwin.

Hidden, A. 1989. *Investigation into the Clapham Junction railway accident* [Department of Transport]. London: HMSO.

Hood, C. C. 1976. *The limits of administration*. London: John Wiley.

— 1989. Introducing politics. In *Politics in Australia*, R. Smith & A. Watson (eds), 3–13. Sydney: Allen & Unwin.

— 1994. *Explaining economic policy reversals*. Buckingham: Open University Press.

Hood, C. C. & G-F. Schuppert (eds) 1988. *Delivering public services in western Europe*. London: Sage.

Hood, C. C, D. K. C. Jones, N. F. Pidgeon, B. A. Turner, R. Gibson, 1992. Risk management. In *Risk: analysis, perception and management*, Royal Society Study Group, 135–92. London: The Royal Society.

Hoover, E. P., R. L. Masters, N. B. Kowalsky 1982. *Eight year review of an employee assistance program for professional pilots*. Denver, Colorado: USALPA.

Horgan, J. 1992. The intellectual warrior. *Scientific American* **267** (November), 20–21.

Horlick-Jones, T. 1990. *Acts of God? an investigation into disasters*. London: EPICENTRE.

— 1991. The nature of disasters. In *Emergency planning in the 90s*, A. Z. Keller & H. C. Wilson (eds), 21–49. Letchworth: British Library/Technical Communications.

— 1995. Modern disasters as outrage and betrayal. *International Journal of Mass Emergencies and Disasters* **13**, 305–15.

Horlick-Jones, T. & G. Peters 1990. Measuring disaster trends, part one: some observations on the Bradford fatality scale. *Disaster Management* 3(3), 144–8.

Horlick-Jones, T., J. Fortune, G. Peters 1991. Measuring disaster trends part two: statistics and underlying processes. *Disaster Management* 4(1), 41–8.

— 1993. Vulnerable systems, failure and disaster. In *Systems science addressing global issues*, F. Stowell, D. West, J. Howell (eds), 559–64. New York: Plenum.

Horlick-Jones, T. & B. De Marchi 1995. The crisis of scientific expertise in *fin de siècle* Europe. In *Scientific expertise in Europe*, T. Horlick-Jones & B. De Marchi (eds), 139–45. *Science and Public Policy* **22**(3) [special issue].

HSE 1978. Canvey: an investigation of potential hazards from operations in the Canvey Island/ Thurrock area. London: HMSO.

— 1981. Canvey: a second report. London: HMSO.

— 1988a. *Blackspot construction: a study of five years' fatal accidents in the building and civil engineering industries*. London: HMSO.

— 1988b. *The tolerability of risk from nuclear power stations*. London: HMSO.

— 1989a. Human factors in industrial safety. HS(G)48. London: HMSO.

— 1989b. *Quantified risk assessment: its input to decision-making*. London: HMSO.

— 1990a. *Safety pays*. London: Health & Safety Executive.

— 1990b. *Human error in risk assessment*. London: AEA Safety and Reliability Directorate/ HSE.

— 1991. *Railway safety: report on the safety report of the railways in Great Britain during 1990.* London: HMSO.

— 1992a. *The tolerability of risk from nuclear power stations*, revised edn. London: HMSO.

— 1992b. Minister warns businesses about "hidden" costs of accidents as UK workplace health and safety week opens. HSE News Release, 23 November.

— 1992c. *Appleton Inquiry report*. London: HMSO.

— 1992d. *Health and Safety Information Bulletin 1992. Safety on the Railways* (195: March), 8–11.

— 1992e. Management of Health and Safety at Work regulations, 1992 and associated Approved Code of Practice [Health and Safety Commission]. London: HMSO.

— 1993. The accident bill – new HSE document adds it up. HSE News Release, 25 January.

— 1995. *Generic terms and concepts in the assessment and regulation of industrial risks* [discussion document]. London: HSE.

235

House of Lords 1993. Regulation of the UK biotechnology industry and global competitiveness [Paper 80: Seventh report of the Select Committee on Science and Technology]. London: HMSO.
Human Factors Revisited, 1993. Editorial. *Airworthy Aviator* **4** (2), 2.

Independent 1990a. Train drivers to fight Purley crash jailing. (6 September)
— 1990b. Railway danger alarms may fall on deaf ears, experts say. (4 September)
— 1990c. Train driver jailed over death crash. (4 September)
— 1991. Pilots hold back safety information. (22 September)
— 1992a. Blaze knocks chemical firm's shares. (1 August)
— 1992b. Fire risk on Tube "does not justify cost of changes". (9 September)
Institute of Electrical and Electronic Engineers. *Spectrum* **24**(2).
International Civil Aviation Organisation 1993. *Human Factors Digest 10: human factors, management and organisation*. Circular 247-AN/148. Montreal, Canada: ICAO.

Jackall, R. 1988. *Moral mazes: the world of corporate managers*. New York: Oxford University Press.
Janis, I. 1972. *Victims of groupthink*. Boston: Houghton–Mifflin.
Janis, I. L. 1971. Groupthink. *Psychology Today* (November), 335–43.
Jenkins, A. 1990. Management at risk. In *11th Advances in Reliability Technology Symposium*, P. Comer (ed.), 59–74. London: Elsevier Applied Science.
Johnston, A. N. 1985. Occupational stress and the airline pilot: the role of the pilot advisory group (PAG). *Aviation, Space, Environmental Medicine* **56**, 633–7.
— 1991. Organisational factors in human factors accident investigation. In *Proceedings of the Sixth International Symposium on Aviation Psychology*, 668–73. Department of Aviation, Ohio State University.
Johnston, A. N. & M. G. Kelly 1988. Post-accident/incident counselling: some exploratory findings. *Aviation, Space, Environmental Medicine* **59**, 766–9.
Jones, D. K. C. 1991. Environmental hazards. In *Global change and challenge*, R. J. Bennett & R. C. Estall (eds), 27–56. London: Routledge.
— 1992. Landslide hazard assessment in the context of development. In McCall (1992: 117–41).
— 1993. Environmental hazards in the 1990s: problems, paradigms and prospects. *Geography* **78**(339), 161–5.
— 1995. Landslide hazard assessment. In *Landslides hazard mitigation*, 96–113. London: The Royal Academy of Engineering.
Jones, R. W. 1973. *Principles of biological regulation: an introduction to feedback systems*. New York: Academic Press.
Jones-Lee, M. W. 1990. The value of transport safety. *The Oxford Review of Economic Policy* **6**(2), 39–60.

Kasperson, R. E. & P. J. M. Stallen (eds) 1990. *Communicating risks to the public*. Dordrecht: Kluwer.
Kates, R. W. 1985. Success, strain and surprise. *Issues in Science and Technology* **2**(1), 46–58.
Kates, R. W. & G. F. White 1961. Flood hazard evaluation. In *Papers on flood problems*, G. F. White (ed.), 135–47. Research Paper 70, Department of Geography, University of Chicago.
Kemp, R. 1991. Risk tolerance and safety management. *Reliability Engineering and System Safety* **31**, 345–53.
— 1993. Risk perception: the assessment of risks by experts and laypeople – a rational comparison?. In *Risk is a construct*, Bayerische Rück (ed.), 103–18. Münich: Knesebeck.
Kharbanda, O. P. & E. A. Stallworthy 1986. *Management disasters*. Cambridge: Gower.
Kingdon, J. 1984. *Agendas, alternatives and public choices*. Boston: Little, Brown.

Kirby, A. M. 1988. High level nuclear waste transportation: political implications of the weakest link in the nuclear fuel cycle. *Environment and Planning C* **6**, 311–22.

Kleindorfer, P., H. Kunreuther, P. Schoemaker 1993. *Decision sciences: an integrative approach*. Cambridge: Cambridge University Press.

Kletz, T. A. 1986. *HAZOP and HAZAN: notes on the identification and assessment of hazards*. London: Institution of Chemical Engineers.

Kloman, H. F. 1990. Risk management agonisties. *Risk Analysis* **10**(2), 201–205.

Knights, D. & T. Vurdubakis 1993. Calculations of risk: towards an understanding of insurance as a moral and political technology. *Accounting, Organizations and Society* **18**(7–8), 729–64.

Kreimer, A. & M. Munasinghe (eds) 1991. *Managing natural disasters and the environment*. Washington: The World Bank.

Kreps, G. 1989. *Social structure and disaster*. Newark: University of Delaware Press.

— 1992. Foundations and principles of emergency planning and management. In *Hazard management and emergency planning: perspectives on Britain*, D. J. Parker & J. Handmer (eds), 159–74. London: James & James.

Krijnen, H. 1979. The flexible firm. *Long Range Planning* **12**, 63–75.

Krimsky, S. & D. Golding (eds) 1992. *Social theories of risk*. New York: Praeger.

Kuhn, T. S. 1962. *The structure of scientific revolutions*. Chicago: University of Chicago Press.

Lacey, N. 1995. Law, politics and criminalisation. In *Frontiers of criminality*, I. Loveland (ed.), 1–27. London: Sweet & Maxwell.

Laird, F. N. 1989. The decline of deference: the political context of risk communication. *Risk Analysis* **9**, 543–50.

LaPiere, T. R. 1934. Attitudes vs actions. *Social Forces* **13**, 230–37.

LaPorte, T. 1982. On the design and management of nearly error-free organizational control systems. In *Accident at Three Mile Island: the human dimensions*, D. Sills (ed), 185–200. Boulder, Colorado: Westview Press.

Latour, B. 1987. *Science in action*. Milton Keynes: Open University Press.

Lautman, L. G. & P. L. Gallimore 1987. Control of the crew caused accident. *Boeing Airliner* (April–June). Seattle: Boeing Commercial Aircraft Company.

Lave, L. B. & E. H. Malès 1989. At risk: the framework for regulating toxic substances. *Environmental Science and Technology* **23**, 386–91.

Layfield, F. 1987. *Sizewell B Public Inquiry*. London: HMSO.

Lechat, M. F. 1990. The International Decade for Natural Disaster Reduction: background and objectives. *Disasters* **14**, 1–6.

Lederer, J. F. 1979. Pros and cons of punishment for achieving discipline in aviation. *ISASI Forum*, Winter 1979.

Lee, K. N. 1993. *Compass and the gyroscope: integrating science and politics for the environment*. New York: Island Press.

Lee, T. R. 1981. The public's perception of risk and the question of irrationality. *Proceedings of the Royal Society of London* **376**, 5–16.

Lewis, H., R. J. Budnitz, H. J. C. Kouts, F. von Hippel, W. Lowenstein, F. Zachariasen 1978. *Risk Assessment Review Group Report to the US Nuclear Regulatory Commission*. Los Angeles, California: Nuclear Regulatory Commission.

Lindblom, C. E. 1959. The science of muddling through. *Public Administration Review* **39**, 79–99.

— 1965. *The intelligence of democracy*. New York: Free Press.

Lipset, S. M. & W. Schneider 1987. *The confidence gap: business, labour and the government in the public mind*. London: Macmillan.

Lloyd, E. & W. Tye 1982. *Systematic safety*. London: Civil Aviation Authority.

Lloyds List 1994. Massive improvement in Shell tanker safety record. (25th January)

Lukes, S. 1974. *Power: a radical view*. Basingstoke: Macmillan.

Mackerron, G. 1983. A case not proven. *New Scientist* **97**(1340), 76–9.

MacNaughton, I. 1977. The price of safety. *Proceedings of the Institution of Civil Engineers* **191**(1), 1–9.

MacPherson, M. 1984. *The black box*. London: Panther Press.

Mahon, Mr Justice 1981. *Report of the Royal Commission to inquire into the crash on Mount Erebus, Antarctica of a DC-10 aircraft operated by Air New Zealand Limited*. Wellington: by authority P. D. Hasselberg, government printer.

Majone, G. 1989. *Evidence, argument and persuasion in the policy process*. New Haven, Connecticut: Yale University Press.

Mangham, I. 1979. *The politics of organisational change*. London: Associated Business Press.

March, J. G. & H. A. Simon 1958. *Organizations*. New York: John Wiley.

Marin, A. 1992. Cost and benefits of risk reduction. In *Risk: analysis, perception and management*, Royal Society Study Group, 192–201. London: Royal Society.

Marschak, T. & R. Nelson 1962. Flexibility, uncertainty and economic theory. *Metroeconomica* **14**, 42–58.

Matza, D. & G. M. Sykes 1961. Juvenile delinquency and subterranean values. *American Sociological Review* **26**(5), 712–19.

Maurino, D. E., J. Reason, N. Johnston, R. B. Lee 1995. *Beyond aviation human factors*. Aldershot: Avebury Aviation.

May, P. J. 1991. Reconsidering policy design: politics and publics. *Journal of Public Policy* **12**(4), 331–54.

McCall, G. J. H., D. J. C. Laming, S. C. Scott (eds) 1992. *Geohazards: natural and man-made*. London: Chapman & Hall.

McGoogan, E. 1984. The autopsy and clinical diagnosis. *Journal of the Royal College of Physicians of London* **18**, 240–3.

Miller, G. A. 1962. *Psychology: the science of mental life*. London: Penguin (Pelican).

Milliken, F. 1987. Three types of uncertainty. *Academy of Management Review* **12**, 133–43.

Minard, R. A. 1993. *Comparative risk: adding value to science*. Northeast Center for Comparative risk Vermont Law School South Royalton, Vermont.

Mintzberg, H. 1979. *The structure of organizations*. Englewood Cliffs, New Jersey: Prentice-Hall.

Mitchell, J. K. 1990. Human dimensions of environmental hazards. In *Nothing to fear*, A. Kirby (ed.), 131–75. Tucson: University of Arizona Press.

Morone, J. & E. Woodhouse 1986. *Averting catastrophe: strategies for regulating risky technologies*. Berkeley: University of California Press.

Morone, J. & E. Woodhouse 1989. *The demise of nuclear energy?* New Haven, Connecticut: Yale University Press.

Morris, T. P. 1963. *Pentonville*. London: Routledge & Kegan Paul.

Munson, R. 1995. Risk associated with, and liability arising from, releases of genetically modified organisms into the environment. *Science and Public Policy* **22**(1), 51–63.

National Research Council 1983. *Risk assessment in the federal government: managing the process*. Washington DC: National Academy Press.

—1989. *Improving risk communication*. Washington DC: National Academy Press.

Needham, M. 1992. Human factors onshore and offshore. In *Major hazards onshore and offshore*, Institution of Civil Engineers, 319–28. Symposium Series 130, Institution of Civil Engineers, Rugby.

Nelson, R. & S. Winter 1982. *An evolutionary theory of economic change*. Cambridge, Massachusetts: Harvard University Press.

New Zealand Treasury 1987. *Government management: brief to the incoming government*, vol. 1. Wellington: Government Printing Office.

Norrie, A. 1991. A critique of criminal causation. *Modern Law Review* **5**, 685–701.

O'Keefe, P., K. Westgate, B. Wisner 1976. Taking the naturalness out of natural disasters. *Nature* **260**(15 April), 566–7.

OECD 1989. *Water resource management: integrated policies*. Paris: Organisation for Economic Cooperation and Development.

O'Neill, J. 1993. *Ecology, policy and politics: human wellbeing and the natural world*. London: Routledge.

O'Riordan, T. 1990. Hazard and risk in the modern world: political models for programme design. In *Hazards and the communication of risk*, J. Handmer & E. C. Penning-Rowsell (eds), 293–301. Aldershot: Gower.

— 1987. Assessing and managing nuclear risk in the United Kingdom. In *Nuclear risk analysis in comparative perspective*, R. Kasperson & J. Kasperson (eds), 197–218. London: Allen & Unwin.

O'Riordan, T., R. Kemp, H. Purdue 1987. On weighing gains and investments at the margin of risk regulation. *Risk Analysis* **7**(3), 361–9.

O'Riordan, T. & B. Wynne, 1987. Regulating environmental risks: a comparative perspective. In *Insuring and managing hazardous risks: from Seveso to Bhopal and beyond*, P. R. Kleindorfer & H. C. Kunreuther (eds), 389–410. Berlin: Springer.

O'Riordan, T. & S. Rayner 1992. Chasing a spectre: risk management and global environmental change. *Global Environmental Change* **1**(2), 95–108.

O'Riordan, T. & J. Cameron (eds) 1994. *Interpreting the precautionary principle*. London: Cameron & May.

Orlady, H. W. 1993. Airline pilot training today and tomorrow. In *Cockpit resource management*, E. L. Wiener, B. G. Kanki, R. L. Helmreich (eds). San Diego: Academic Press.

Ostrom, V. 1974. *The intellectual crisis in American public administration*, revised edn. Tuscaloosa: Alabama University Press.

Otway, H. & P. Pahner 1980. Risk assessment. In *Risk and chance: selected readings*, J. Dowie & P. Lefrere (eds), 140–68. Milton Keynes: Open University Press.

Ozonoff, D. 1993. Science and the grassroots toxics movement: ten years of partnership, everyone's back yard. *Citizens' Clearinghouse for Hazardous Waste* (July/August), 18–19.

Palm, R. I. 1990. *Natural hazards: an integrated framework for research and planning*. Baltimore: Johns Hopkins University Press.

Palm, R. I., M. Hodgson, R. Blanchard, D. Lyons 1990. *Earthquake insurance in California*. Boulder: Westview Press.

Parker, D. J. 1992. Flood disasters in Britain: lessons from flood hazard research. *Disaster prevention and management* **1**(1), 8–25.

— 1995. Floodplain development policy in England and Wales. *Applied Geography* **15**(4), 341–63.

Parker, D. J. & E. C. Penning-Rowsell 1983. Flood hazard research in Britain. *Progress in Human Geography* **7**(2), 182–202.

Parker, D. J. & J. Handmer (eds) 1992. *Hazard management and emergency planning: perspectives on Britain*. London: James & James.

Peltzman, S. 1989. The economic theory of regulation after a decade of deregulation. *Brookings Papers on Economic Activity (Microeconomics)* 1–41.

Penning-Rowsell, E. C. 1996 (in press). Flood hazard response in Argentina: changing context and changing policies. *Geographical Review* **86**.

Penning-Rowsell, E. C., D. J. Parker, D. Crease, C. R. Mattison 1983. *Flood warning dissemination: an evaluation of some current practices in the Severn Trent Water Authority area*. Middlesex Polytechnic School of Planning Studies Papers 7, School of Geography and Planning, Middlesex University.

Penning-Rowsell, E. C., D. J. Parker, D. M. Harding 1986. *Floods and drainage: British policies for hazard reduction, agricultural improvement and wetland conservation*. London: Allen & Unwin.

Penning-Rowsell, E. C. & P. Winchester 1992. Scenario construction for risk communication in emergency planning: six "golden rules". In *Hazard management and emergency planning: perspectives on Britain*, D. J. Parker & J. Handmer (eds), 203–218. London: James & James.

Penning-Rowsell, E. C., C. H. Green, P. M. Thompson, A. C. Coker, S. M. Tunstall, C. Richards, D. J. Parker 1992a. *The economics of coastal management: a manual of assessment techniques*. London: Pinter (Belhaven).

Penning-Rowsell, E. C. et al. 1992b. Flood vulnerability analysis and climatic change: towards a European methodology. In *Floods and flood management*, A. J. Saul (ed.), 343–61. London: Kluwer.

Penning-Rowsell, E. C. & M. Fordham 1994. *Floods across Europe: hazard assessment, modelling and management*. London: Middlesex University Press.

Penning-Rowsell, E. C. & S. M. Tunstall 1996 (in press). Risks and resources: defining and managing the floodplain. In *Floodplain processes*, M. Anderson, D. Walling, P. Bates (eds). Chichester: John Wiley.

Perrow, C. 1984. *Normal accidents: living with high-risk technologies*. New York: Basic Books.

Peters, T. & R. Waterman 1982. *In search of excellence*. New York: Harper & Row.

Petts, J. 1994. Effective waste management: understanding and dealing with public concerns. *Waste Management and Research* 12, 207–222.

Pidgeon, N. F. 1988. Risk assessment and accident analysis. *Acta Psychologica* 68, 355–68.

Pidgeon, N. F., C. Hood, D. K. C. Jones, B. A. Turner, R. Gibson 1992. Risk perception. In *Risk: analysis, perception and management*, Royal Society Study Group, 89–134. London: Royal Society.

Pidgeon, N. F. & M. O'Leary 1994. Organizational safety culture: implications for aviation practice. In *Aviation psychology in practice*, N. Johnston, N. McDonald, R. Fuller (eds), 21–43. Aldershot: Avebury Technical.

Pilisuk, M., S. H. Parks, G. Hawkes 1987. Public perceptions of technological risk. *The Social Science Journal* 24(4), 403–413.

Piller, C. 1991. *The fail safe society: community defiance and the end of American technological optimism*. New York: Basic Books.

Pitblado, R. M. & D. H. Slater 1990. *Quantitative assessment of process safety programs*. Internal company document. Technica Inc., 335 East Campus View Blvd, Columbus, OH43085.

Polisar, D. & A. Wildavsky 1989. From individual to system blame: a cultural analysis of change in the law of torts. *Journal of Policy History* 1(2), 129–55.

Posner, R. A. 1986. *Economic analysis of law*, 3rd edn. Boston: Little Brown.

Power, M. 1994. *The audit explosion*. London: DEMOS.

Quarantelli, E. L. 1978. *Disasters, theory and research*. Beverly Hills, California: Sage.

Quinn, J. 1980. *Strategies for change: logical incrementalism*. Homewood, Illinois: Irwin.

Rabin, R. 1992. A socio-legal history of the tobacco tort litigation. *Stanford Law Review* 44, 853.

Raiffa, H. 1968. *Decision analysis*. Reading, Massachusetts: Addison-Wesley.

Rasmussen, J. 1990. Human error and the problem of causality in analysis of accidents. *Philosophical Transactions of the Royal Society of London* 327B, 449–62.

Ravetz, J. R. 1971. *Scientific knowledge and its social problems*. Oxford: Oxford University Press.

—1990. *The merger of knowledge with power: essays in critical science*. London: Mansell.

Rayner, S. 1992. Cultural theory and risk analysis. In *Social theories of risk*, J. Krimsky & D. Golding (eds), 83–117. New York: Praeger.

—1989. Risk, uncertainty and social organization. *Contemporary Sociology* 18, 6–9.

RCEP (Royal Commission on Environmental Pollution) 1988. *Best practicable environmental option* [12th Report]. London: HMSO.

— 1989. *The release of genetically engineered organisms to the environment* [13th report]. London: HMSO.

— 1991. GENHAZ: *A system for the critical appraisal of proposals to release genetically modified organisms into the environment* [14th Report]. London: HMSO.

Reason, J. 1990. *Human error.* New York: Cambridge University Press.

— 1991. Too little and too late: a commentary on accident and incident report systems. In *Near miss reporting as a safety tool,* D. A. Lucas & A. R. Hale (eds), 109–20. London: Butterworth Heinmann.

Reid, S. G. 1992. Acceptable risk. In *Engineering safety,* D. Blockley (ed.), 138–66. London: McGraw-Hill.

Renn, O. & D. Levine 1992 Trust and credibility. In *Risk communication,* H. Jungermann, R. E. Kasperson, P. M. Wiedemann (eds). KPA, Julich.

Reuter, J. 1988. *The economic consequences of expanded corporate liability: an exploratory study.* Santa Monica: Rand.

Reynard, W. D., C. E. Billings, E. S. Cheaney, R. Hardy, *The development of the NASA Aviation Safety Reporting System.* Reference Publication 1114 National Aeronautics and Space Administration, Moffet Field, California.

Riker, W. H. 1962. *The theory of political coalitions.* New Haven, Connecticut: Yale University Press.

Rimington, J. D. Overview of risk assessment. Paper presented to the International Conference on Risk Assessment, organized by HSE and jointly sponsored by HSC, the Commission of the European Communities, the International Labour Organisation, and the Organisation for Economic Cooperation and Development, London, October 1992.

— 1993. Coping with technological risk: a 21st century problem. CSE Lecture, Royal Academy of Engineering, London.

Roberts, K. H. 1989. New challenges in organizational research: high reliability organizations. *Industrial Crisis Quarterly* 3(2), 111–25.

Roberts, K. H. & G. Gargano 1989. Managing a high reliability organization: a case for interdependence. In *Managing complexity in high technology organizations: systems and people,* M. A. Slinow & S. Mohrman (eds), 146–59. New York: Oxford University Press.

Rochlin, G. I. 1989. Informal organizational networking as a crisis-avoidance strategy: US naval flight operations as a case study. *Industrial Crisis Quarterly* 3(2), 159–76.

Rosenhead, J., J. Elton, S. Gupta 1972. Robustness and optimality as criteria for strategic decisions. *Operational Research Quarterly* 23, 413–31.

Royal Society 1983. *Risk assessment: a study group report.* London: The Royal Society.

Royal Society Study Group 1992. *Risk: analysis, perception, management.* The Royal Society, London.

Russell, D. 1993 *The Earth, humanity and God.* London: UCL Press.

Ryan, W. 1976. *Blaming the victim.* New York: Vintage books.

Sachdeva, P. 1984. Development planning an adaptive approach. *Long Range Planning* 17, 96–102.

Sagan, S. 1993. *The limits of safety: organizations, accidents and nuclear weapons.* Princeton, New Jersey: Princeton University Press.

Sawhill, J. & L. Silverman 1983. Build flexibility not power plants. *Public Utilities Fortnightly* 111, 96–102.

Scanlon, J. 1988. Winners and losers: some thoughts about the political economy of disaster. *International Journal of Mass Emergencies and Disasters* 6, 47–64.

Schlesinger, J. R. 1967. *Systems analysis and the political process* [mimeograph]. Washington DC: RAND Corporation.

Schon, D. 1983. *The reflective practitioner.* London: Temple Smith.

Schulmeyer, G. 1990. *Zero defect software.* New York: McGraw-Hill.

Schwarz, M. & M. Thompson 1990. *Divided we stand: redefining politics, technology and social choice*. Hemel Hempstead: Harvester Wheatsheaf.

Seatrade Review 1994. Legs to stand on. (29–31 March)

Senge, P. M. 1990. *The fifth discipline: the art and practice of the learning organisation.* London: Century Business Books.

Sewell, W. R. D. & H. D. Foster 1976. *Images of Canadian futures: the role of conservation and renewable energy.* Ottawa: Fisheries and Environment Canada.

Sheen, Mr Justice 1987. *MV Herald of Free Enterprise* [Report of Court 8074 Formal Investigation: the *"Sheen Report"*]. London: Department of Transport.

Shell International 1988. *Shell Safety Management Program.* The Hague: Shell International.

Shkler, J. 1964. Decisionism. In *Nomos* (vol. 17), C. J. Friedrich (ed.), 3–17. New York: Atherton.

Shrader-Frechette, K. S. 1985. *Risk analysis and scientific method.* Dordrecht: Reidel.

— 1991a. *Risk and rationality: philosophical foundations for populist reforms.* Berkeley: University of California Press.

— 1991b. Reductionist approaches to risk. In *Acceptable evidence: science and values in risk management*, D. G. Mayo & R. D. Hollander (eds), 218–48. New York: Oxford University Press.

Sieber, S. 1981. *Fatal remedies: the ironics of social intervention.* New York: Plenum.

Sime, J. D. 1985. Designing for people or ball-bearings? *Design Studies* 6(3), 163–8.

Slatter, S. 1984. *Corporate recovery.* London: Penguin.

Slovic, P. 1992. Perception of risk: reflections on the psychometric paradigm. In *Social theories of risk*, S. Krimsky & D. Golding (eds), 117–52. West Point, Connecticut: Praeger.

Smith, K. 1992. *Environmental hazards.* London: Routledge.

— 1995. *Environmental hazards*, 2nd edn. London: Routledge.

Spencer, J. R. 1985. Motor vehicles as weapons of offence. *Criminal Law Review* (January), 29–41.

Spooner, P. 1992. Corporate responsibility in an age of deregulation. In *Hazard management and emergency planning*, D. Parker & J. Handmer (eds), 95–108. London: James & James.

Starr, C. 1969. Social benefit versus technological risk. *Science* **165**, 1232–38.

Stern, P. C. 1991. Learning through conflict: a realistic strategy for risk communication. *Policy Sciences* **24**, 99–119.

Stigler, G. 1971. The theory of economic regulation. *Bell Journal of Economics and Management Science* 2(3), 3–21. [Reprinted in 1988 in *Chicago studies in political economy*, G. Stigler (ed.), 209–33. Chicago: University of Chicago Press.]

Stone J. R, D. I. Blockley, B. W. Pilsworth 1989. Towards machine learning from case histories. *Journal of Civil Engineering Systems* 6(3), 129–35.

Strutt, J. E. & K. Allsopp 1993. Corrosion risk analysis: an inspection and maintenance policy. *Proceedings of the Institution of Mechanical Engineers*, paper C446/067/93, 87–99.

Sunday Times 1989. It's the doers wot get the blame. (20 August)

Susman, P., P. O'Keefe, B. Wisner 1983. Global disasters, a radical interpretation. In *Interpretations of calamity*, K. Hewitt (ed.), 263–83. London: Allen & Unwin.

Tait, E. J. & L. Levidow 1992. Proactive and reactive approaches to risk regulation: the case of biotechnology. *Futures* **24** (April), 219–31.

— 1993. Advice on biotechnology regulation: the remit and composition of Britain's ACRE. *Science and Public Policy* **20**(3), 193–209

Tajfel, H. 1981. *Human groups and social categories.* Cambridge: Cambridge University Press.

Taylor, I. 1983. *Crime, capitalism and community.* London: Butterworth.

Thomas, K. 1971. *Religion and the decline of magic.* London: Weidenfeld & Nicolson.

Thomas, S. 1988. *The realities of nuclear power: international economic and regulatory experience.* Cambridge: Cambridge University Press.

Thompson, M., R. Ellis, A. Wildavsky 1990. *Cultural theory.* Boulder, Colorado: Westview.

BIBLIOGRAPHY

The Times 1990. Safety work in N. Sea "will cut oil production". (2 February)

Toft, B. 1984. *Human factor failure in complex systems*. Dissertation, School of Independent Studies, University of Lancaster.

— 1990. *The failure of hindsight*. PhD thesis, Department of Sociology, University of Exeter.

Toft, B. & S. Reynolds 1994. *Learning from disasters: a management approach*. Oxford: Butterworth–Heinemann.

Torry, W. I. 1978. Natural disasters, social structure and change in traditional societies. *Journal of African and Asian Studies* **13**, 167–83.

Turner, B. A. 1978. *Man-made disasters*. London: Wykeham Press.

— 1991. The development of a safety culture. *Chemistry and Industry* (1 April), 241–3.

— 1994a. The future of risk research. *Journal of Contingencies and Crisis Management* **2**(3), 146–56.

— 1994b. Causes of disaster: sloppy management. *British Journal of Management* **5**, 215–19.

Turner, B. A., N. Pidgeon, D. Blockley, B. Toft 1989. Safety culture: its importance in future risk management. Position paper for the Second World Bank Workshop on Safety Control and Risk Management, 6–9 November, Karlstad, Sweden.

UNDRO 1990. World launches International Decade for Natural Disaster Reduction. *UNDRO News* (January/February).

US Environmental Protection Agency 1992. *Environmental equity: reducing risk for all communities*. Washington DC: USEPA.

Vette, G. (with J. Macdonald) 1983. *Impact Erebus*. Auckland: Hodder & Stoughton.

Wagennar, W. & J. Groenewold 1987. Accidents at sea; multiple causes and impossible consequences. *International Journal of Man–Machine Studies* **27**, 587–98.

Waring, A. 1992. Developing a safety culture. *The Safety & Health Practitioner* **10**(4), 42–4.

Warner, F. 1992. Introduction. In Royal Society (1992: 1–12).

— 1993. Calculated risks. *Science Policy Analysis*, 44–9.

Watson, S. R. & D. M. Buede 1987. *Decision synthesis: the principles and practice of decision analysis*. Cambridge: Cambridge University Press.

Weale, A. 1992. *The new politics of pollution*. Manchester: Manchester University Press.

Weik, K. 1989. Mental models of high reliability systems. *Industrial Crisis Quarterly* **3**(2), 127–42.

Weinberg, A. M. 1972. Science and trans-science. *Minerva* **10**, 209–22.

Weir, D. T. H. 1975. *Stress and the manager in the over-controlled organization*. In *Managerial stress*, D. Gowler & K. Legge (eds), 165–78. Epping: Gower.

— 1993. Communication factors in system failure or why big planes crash and big businesses fail. *Disaster Prevention and Management* **2**(2), 41–50.

Wells, C. 1988. The decline and rise of English murder: corporate crime and individual responsibility. *Criminal Law Review* (December), 788–801.

— 1991. Inquests, inquiries and indictments: the official reception of death by disaster. *Legal Studies* **11**(1), 71–84.

— 1993a. *Corporations and criminal responsibility*. Oxford: Oxford University Press.

— 1993b. Disasters: the role of institutional responses in shaping public perceptions of death. In *Death rites*, R. Lee & D. Morgan (eds), 196–222. London: Routledge.

— 1995a. *Negotiating tragedy: law and disasters*. London: Sweet & Maxwell.

— 1995b. Corporate manslaughter: a cultural and legal form. *Criminal Law Forum* **6**, 45–72.

— 1995c. A quiet revolution in corporate liability for crime. *New Law Journal* **145**, 1326–7.

Westrum, R. 1992. Cultures with requisite imagination. NATO Verification and Validation Conference, Vimeiro, Portugal.

— 1993. Cultures with requisite imagination. In *Verification and validation in complex man-machine systems*. J. Wise, P. Stager, J. Hopkin (eds). New York: Springer.

Wheale, P. & R. MacNally (eds). *The bio-revolution: cornucopia or Pandora's Box?* London: Pluto.

Wiener E. L. & D. C. Nagel (eds) 1988. *Human factors in aviation*. San Diego: Academic Press.

Wiener E. L., B. G. Kanki, R. L. Helmreich (eds) 1993. *Cockpit resource management*. San Diego: Academic Press.

Wijkman, A. & L. Timberlake 1984. *Natural disasters: acts of God or acts of man?* London: Earthscan.

Wildavsky, A. 1985. *Trial without error: anticipation versus resilience as strategies for risk reduction*. Sydney: Centre for Independent Studies.

—1988. *Searching for safety*. New Brunswick: Transaction Books.

—1971. *The revolt against the masses*. New York: Basic Books.

Wilkins, L. & P. Patterson 1990. The political amplification of risk: media coverage of disasters and hazards. In *Hazards and the communication of risk*, J. Handmer & E. Penning-Rowsell (eds), 79–94. Aldershot: Gower Technical.

Williams, R. 1980. *The nuclear power decisions*. London: Croom Helm.

Wilpert, B. 1991. System safety and safety culture. Paper presented at joint meeting of IAEA/IIASA: The Influence of Organization and Management on the Safety of NPPs and other Industrial Systems.

Wilson, J. Q. 1980. The politics of regulation. In *The politics of regulation*, J. Q. Wilson (ed.), 357–74. New York: Basic Books.

von Winterfeldt, D. & W. Edwards 1984. Patterns of conflict about risky technologies. *Risk Analysis* **4**, 55–68.

Wolford, V. L. 1960. *Tornado occurrence in the United States*. Washington DC: US Department of Commerce, Weather Bureau.

Wright, J. D. 1988. Second thoughts about gun control. *The Public Interest* **91**(Spring 1988), 23–39.

Wynne, B. (ed.) 1987. *Risk management and hazardous waste: implementation and the dialectics of credibility*. Berlin: Springer.

—1992a. Risk and social learning: reification to engagement. In *Social theories of risk*, S. Krimsky & D. Golding (eds), 275–97. Westport, Connecticut: Praeger.

—1992b. Uncertainty and environmental learning: reconceiving science and policy in the preventive paradigm. *Global Environmental Change* **2**(June), 111–27.

—1992c. Public understanding of science: new horizons or hall of mirrors? *Public Understanding of Science* **1**, 37–43.

—1992d. Risk and social learning: reification to engagement. In *Theories of risk*, S. Krimsky & D. Golding (eds), 275–97. New York: Praeger.

Yalow, R. S. 1985. Radioactivity in the service of humanity. *Thought* [Fordham University Quarterly] **60**, 236.

Zimmerman, R. 1985. Private sector response patterns to risks from chemicals. In *Risk analysis in the private sector*, C. Whipple & V. Covello (eds), 15–32. New York: Plenum.

Index

245

Lightning Source UK Ltd.
Milton Keynes UK
20 January 2011

166007UK00003B/63/A